Minitab Guid
Moore's
The Basic Practice of Statistics
Second Edition

Betsy S. Greenberg
The University of Texas at Austin

W. H. Freeman and Company
New York

Cover image: "M&M's" Chocolate Candies is a registered trademark of Mars, Incorporated and used with permission of the owner. © Mars, Inc.

ISBN 0-7167-3613-6

MINITAB® is registered trademark of Minitab, Inc. Output from MINITAB is printed with permission of Minitab, Inc., State College, PA.

Copyright © 2000 by W. H. Freeman and Company

No part of this book may be reproduced by any mechanical, photographic, or electronic process, or in the form of a phonographic recording, nor may it be stored in a retrieval system, transmitted, or otherwise copied for public or private use, without written permission from the publisher.

Printed in the United States of America

First printing 1999

Table of Contents

Preface v

Introduction to Minitab 1

Chapter 1 - Examining Distributions 13

Chapter 2 - Examining Relationships 45

Chapter 3 - Producing Data 69

Chapter 4 - Probability and Sampling Distributions 85

Chapter 5 - Probability Theory 101

Chapter 6 - Introduction to Inference 111

Chapter 7 - Inference for Distributions 121

Chapter 8 - Inference for Proportions 139

Chapter 9 - Inference for Two-way Tables 151

Chapter 10 - One-way Analysis of Variance 159

Chapter 11 - Inference for Regression 167

Chapter 12 - Nonparametric Tests 183

Appendix - Minitab Commands and Menu Equivalents 195

Index 201

Preface

This *Minitab Guide* accompanies the Second Edition of *The Basic Practice of Statistics* (BPS) by David S. Moore and is intended to be used with the statistical software package called Minitab. Minitab was originally developed in 1972 to help professors teach basic statistics. The software is now used in more than 2000 colleges and universities around the world. Minitab relieves students of tedious statistical calculations and allows them to better understand statistical concepts. Minitab is also the tool of choice for businesses of all sizes. It is used in 80 countries throughout the world; from start-ups to the Fortune 500 companies, including Ford Motor Company, 3M, AlliedSignal, General Motors and Lockheed Martin.

Minitab is available on a wide variety of computers, including mainframes and personal computers. This book is based on Release 12, the most recent version of Minitab available. Both menu and session commands are illustrated. All versions of Minitab allow the session commands. Windows and Macintosh versions also allow menu commands. If you are using a version different from Release 12, there may be slight differences in the menu interfaces. For further information about the software, contact

> Minitab Inc.
> 3081 Enterprise Drive
> State College, PA 16801 USA
> Phone: (814) 238-3280
> Fax: (814) 238-4383
> e-mail: Info@minitab.com
> URL: http://www.minitab.com

This *Minitab Guide* parallels the Second Edition of BPS. The *Minitab Guide* contains an introduction to Minitab plus a chapter corresponding to each chapter in BPS. In each chapter, we show how Minitab can be used to perform the statistical techniques described in BPS. In addition, each chapter includes exercises selected and modified from BPS that are appropriate to be done using Minitab. The numbering of the exercises refers to exercises in BPS. The Appendix lists by topic Minitab commands and menu equivalents that are referred to in this guide.

Minitab worksheets referred to in this guide are available on the W.H. Freeman website (http://www.whfreeman.com/statistics/bps/index.htm). The names of worksheets refer to examples, exercises, and tables in BPS.

Introduction to Minitab

Topics to be covered in this chapter:

What Minitab Will Do for You
Different Versions of Minitab
Beginning and Ending a Minitab Session
The Minitab Worksheet
The Data Window and Entering Data
Minitab Commands
The Info Window
Opening, Saving, and Printing Files
Managing and Calculating Data
Getting Help

What Minitab Will Do for You

Before the widespread availability of powerful computers and prepackaged statistics programs, manual computations were emphasized in statistics courses. Today, computers have revolutionized data analysis, which is a fundamental task of statistics. Packages such as Minitab allow the computer to solve statistics problems. Minitab can perform a wide variety of tasks, from the construction of graphical and numerical summaries for a set of data to the complicated statistical procedures and tests described in the book *Basic Practice of Statistics*, Second Edition, by David S. Moore. Minitab will allow you to concentrate less on the mathematical calculations and more on the analysis of the data. In this supplement, we will refer to the textbook as BPS. The numbering of exercises also refers to exercises in BPS.

Different Versions of Minitab

Minitab runs on Windows and Macintosh computers, as well as most of the leading workstations and mainframe computers. This book is based on Release 12, the most recent version of Minitab available. Different versions may look slightly different on the screen and allow different methods of executing commands. All versions of Minitab allow you to type commands. In addition, Windows and Macintosh versions have menus that allow you to execute commands. This book will illustrate both menu and session commands. If you are using a version different from Release 12, there may be some differences in the menu interfaces. Manuals that come with the software as well as online help are available to give you more information.

Beginning and Ending a Minitab Session

To start a Minitab session from the menu, choose **Start ➤ Programs ➤ Minitab 12 for Windows ➤ Minitab**. To exit Minitab, choose **File ➤ Exit**.

To start Minitab in Windows NT 3.51 or on Macintosh systems, locate the Minitab icon and double click. On mainframe versions, type MINITAB at the prompt for your computer system. A Minitab session can be ended by typing STOP at the Minitab prompt (MTB >).

When you first enter Minitab, the screen will appear as in the figure with a toolbar, a Session window and a Data window. Additional windows such as the Info window, History window, graph windows and dialog boxes may also appear as you use Minitab.

The Minitab Worksheet

The worksheet is arranged by rows and columns. The columns, C1, C2, C3, etc., correspond to the variables in your data, the rows to observations. The columns can be viewed in the Data window. In addition, the worksheet may also include stored constants, K1, K2, K3, etc. Stored constants can be viewed in the Info window.

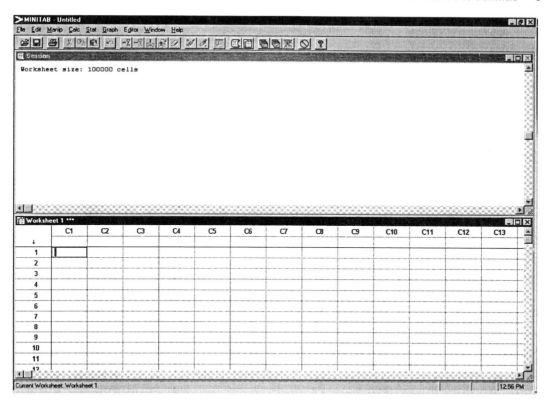

Most of the Minitab commands address the columns. In general, a column contains data for one variable, and each row contains all the data for each subject or observation. Columns can be referred to by number (C1, C2, C3, etc.) or by names such as "height" or "weight".

Constants are referenced by the letter K and a number (K1, K2, K3, etc.). Unlike columns, constants are single values. Storing a constant tells Minitab to remember this value; it will be needed later. Constants are analogous to the memory functions on most calculators. For instance, Minitab allows you to quickly find the average of a column of numbers. Instead of having to write it down, the value can be stored into a constant such as K1 and used in subsequent calculations.

Minitab Commands

Commands tell Minitab what to do. The **STOP** command, tells Minitab to end a session. You can issue commands in Minitab by choosing commands from the menus or by typing session commands directly into the Session window. To issue a command from a menu, click on an item in the menu bar to open the menu, then click on a menu item to execute the command, open a submenu, or open a dialog box. To use the **STOP** command, choose **File ➤ Exit**. For many of the menu commands, a dialog box will appear, prompting you for additional information needed to carry out the command.

Session commands are an alternative to menu commands. Most session commands are simple, easy-to-remember words like **PLOT**, **SAVE**, or **HISTOGRAM**. You can type commands into the Session window by clicking on the Session window and choosing **Editor ➤ Enable Command Language**. If the Editor menu shows Disable Command Language, you don't have to do anything. Type the session commands at the Minitab prompt (MTB >). After each command (or subcommand), press Enter. Session commands are available for all versions of Minitab, including mainframe versions.

Commands always start with a command word, such as **PLOT**. Only the first four letters of a command word are needed. A command may require more data, such as the column(s) and/or constant(s) on which the action is performed. Additional letters and text may be added for clarity. Minitab accepts either upper- or lowercase and does not distinguish between them.

Command syntax is described in this book using special symbols, for example C, K, E, FILENAME, and square brackets. When you use a command, replace C by a specific column, using either the number (e.g., C2) or name, enclosed in single quotes (e.g., 'HEIGHT'). Replace K by a number (e.g., 5.2) or a stored constant (e.g., K4). Replace E by either a column, a number, or a stored constant. Some commands have optional arguments. When the syntax for these commands is given, the optional arguments are enclosed in square brackets. Some commands may refer to files. To use these, replace FILENAME by a specific name, enclosed in single quotes.

Both column names and file names are enclosed in single quotes. Some computer keyboards have both right (') and left (`) single quotes. The right quote must be used; the left quote is not recognized in Minitab.

The Data Window and Entering Data

The Data window shows the columns in your worksheet and allows you to easily enter, edit and view your data. To enter a value in a Data window cell, just click on the cell, type a value, and press Enter. To enter a column of data, click the data direction arrow to make it point down. To enter a row of data, click the data direction arrow to make it point to the right. The data direction arrow can also be changed by choosing **Editor ➤ Change Entry Direction**.

Data direction arrow

	C1-T	C2	C3
	State	Percent	
1	AL	13.0	
2	AK	5.2	
3	AZ	13.2	
4	AR	14.4	
5	CA	10.5	
6	CO	11.0	
7	CN	14.3	
8	DE		
9	FL		
10	GA		

Enter your data, pressing Tab or Enter to move down or across. Press Ctrl + Enter to move to the start of the next column or row. Click on row one, column two, then type 13, Enter, 5.2, Enter, etc. Notice that when you type a value and press Enter, the active cell moves down.

Using the **Edit ➤ Copy** and **Paste** commands, you can copy from a variety of sources and paste to the Data window. You can copy from cells, rows, or columns of the same or another data window. You can also copy from other applications such as spreadsheets or word processors.

The **SET** command is also used to enter data into a Minitab worksheet one column at a time. The format for the command is as follows:

```
SET the following data into C
```

where C designates a single column, for example, C2. After entering the **SET** command, Minitab expects data and returns the data prompt (DATA>). Entered numbers may be separated with either spaces or commas. Each time the return key is used, Minitab will return the data prompt. The command **END** signifies that the data for the column are complete. Minitab will discontinue the data prompt and return to the Minitab prompt.

6 *Introduction to Minitab*

To name a column, click on the column name cell and type the name. Names cannot be longer than 31 characters, begin or end with a space, include the symbol ' or #, or start with the symbol *. Columns can also be named using session commands. The format for the **NAME** command is

```
NAME E = 'name' ... E = 'name'
```

If we want to name the columns in our current worksheet, we might use the names illustrated below.

```
MTB > name c1 'State' c2 'Percent'
```

The INFO Window and Command

This window summarizes all the data in the active worksheet columns, constants, and matrices. You can choose to see the Info window by choosing **Window ➤ Info**. You close the Info window by clicking the close button on the title bar (as you would any window on your system).

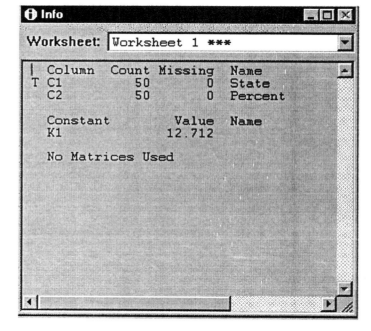

The **INFO** command can be used to list the contents of the current Minitab worksheet in the Session window. The format for the **INFO** command is

```
INFO   [C...C]
```

For the **INFO** command, specifying columns is optional, as indicated by the square brackets. If no columns are specified, **INFO** lists the attributes of all columns and constants. For the current worksheet, the **INFO** command would display the following information:

Introduction to Minitab 7

```
MTB > info
```

Information on the Worksheet

```
     Column   Count   Name
T    C1         50    State
     C2         50    Percent

     Constant        Value   Name
     K1            12.7120
```

The output generated above reveals that there are currently fifty data elements each or observations in column C1 and C2. Constant K1 has the value 12.712 (the mean of C2).

Extra text may be added to Minitab commands. In the following example, the words "on the column" are added to increase the clarity of the command.

```
MTB > info on the column c1
```

Information on the Worksheet

```
     Column   Count   Name
T    C1         50    State
```

Opening, Saving, and Printing Files

To open data from a file, choose **File ➤ Open Worksheet**. In the Files of type box, choose the type of file you are looking for: Minitab, Minitab portable, Excel, etc. Select a file, then click Open.

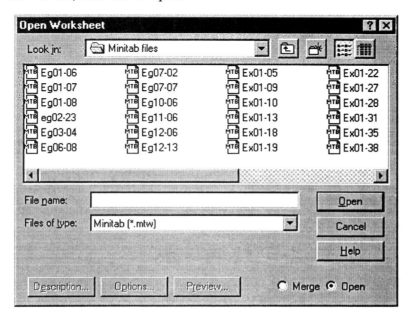

8 *Introduction to Minitab*

To print the contents of the data window, click on the data window and choose **File ➤ Print Worksheet**. Select the desired options in the dialog box and click OK. The contents of the worksheet can also be printed using the `PRINT` command.

The `PRINT` command displays data contained in the worksheet on your screen. `PRINT` will display the values contained in either columns or constants. The format for the command is

```
PRINT E...E
```

If only one column is specified, the data will be displayed horizontally, without row numbers. Minitab will print the data with row numbers if more than one column is specified.

Each data set you work with is contained in a worksheet. In Minitab Version 12, you can have many worksheets and graphs in one project. To save data as part of a project, choose **File ➤ Save Project**. To save data into a separate file, make the desired Data window active, choose **File ➤ Save Worksheet As**. In Save as type, choose the data format in which you want the data to be saved. Select a directory, enter a file name, and click Save.

The session command `SAVE` command allows the data in a worksheet to be stored so that it can be used later. The format for the `SAVE` command is

```
SAVE [in file in "filename" or K]
```

Changing the Data

Minitab provides four session commands to change the values in the data set: `INSERT`, `DELETE`, `LET`, and `ERASE`. The `INSERT` command allows rows to be added at the beginning, middle, or end of a column. The `DELETE` command will remove data items from columns. The `LET` command can be used to change individual data elements. The `ERASE` command is used to remove constants or entire columns from the data set.

The `INSERT` command is used to add data at the top, between two rows, or at the bottom of columns of data. The format for the `INSERT` command is

```
INSERT data [between rows K and K] of C...C
```

The columns C...C designate the columns on which to perform the insertion. The constant values K and K specify between which two rows the data are to be inserted. To insert data at the beginning of a column, the rows zero and one are specified. To insert data in the middle of the columns, use the row numbers between which you want to insert the data. To append data to the end of a column, specify only the columns to which you want to append; do not specify any row numbers.

The **DELETE** command removes unwanted data from columns. The format for the command is:

```
DELETE rows K...K of C...C
```

The constant values K...K designate which row numbers to delete. The columns C...C determine on which column to perform the operation.

The **ERASE** command can be used to remove columns or constants that you no longer need. The format for the command is

```
ERASE E...E
```

The **LET** command can be used to change a single observation in a column. It is commonly used to correct data entry errors. The format for the command is

```
LET C(K) = K
```

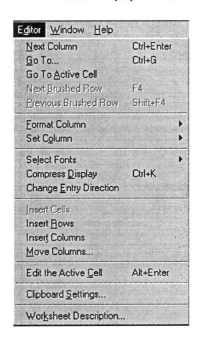

The **LET** command is also used to calculate values using arithmetic. This will be described in Chapter 1.

The Editor menu can be used instead of the **INSERT**, **DELETE**, and **ERASE** commands. Under the Editor menu, you may select to either insert cells, rows, or columns as illustrated to the right. In the Data window, columns, rows, or individual cells can be highlighted and then deleted using the Delete key or by choosing **Edit ► Delete cells** from the menu bar.

Getting HELP

Documentation to explain session commands is available using the **HELP** command. The format for the **HELP** command is

HELP [command]

Help is obtained for the **INFO** command below.
```
MTB > help info
```

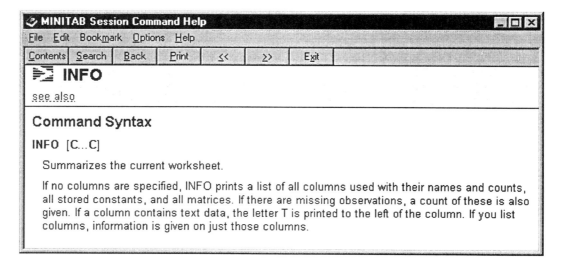

Help is available even for the **HELP** command by typing HELP HELP. The **HELP** command can assist users in becoming adept at Minitab. In fact, **HELP** can be consulted before referencing this guide. The online help is concise and at a user's fingertips. The **HELP** command should be used often to increase Minitab's effectiveness.

Documentation on Minitab features and concepts, written for users of menus and dialog boxes can be obtained by clicking the Help button in any dialog box or pressing F1 at any time. Help can be obtained from the menu by selecting **Help ➤ Search for Help on** and then selecting a topic (such as Info). In this case, Minitab will provide help with the menu command instead of the session command. This is illustrated on the following page.

Introduction to Minitab 11

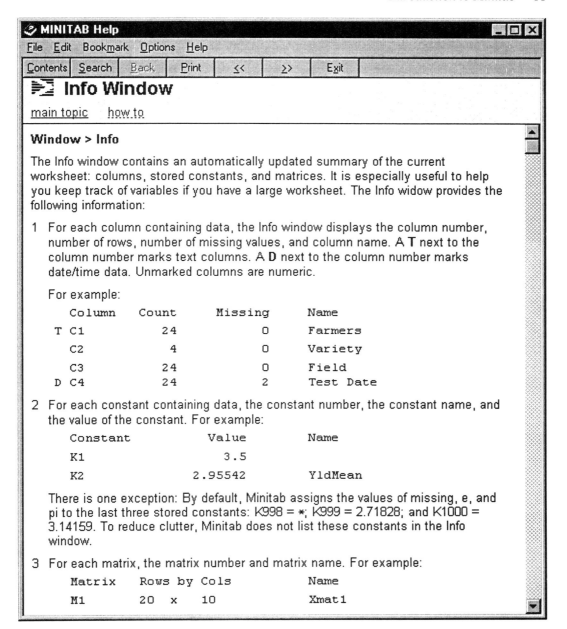

Chapter 1
Examining Distributions

Commands to be covered in this chapter:

```
STEM-AND-LEAF display of C...C
HISTOGRAM of C...C
STACK (E...E) on ... on (E...E), put in (C...C)
UNSTACK (C...C) into (E...E) ... (E...E)
TSPLOT [period = K] of C
DESCRIBE variables in C...C
BOXPLOT of C...C
MEAN of the values in C [put into K]
RMEAN of E...E put into C
CENTER the data in C...C put into C...C
LET = expression
CDF for values in E...E [put results in E...E]
INVCDF for values in E [put into E]
```

Displaying Distributions

The distribution of a variable can be displayed graphically with Minitab commands such as **HISTOGRAM** and **STEM-AND-LEAF**. Histograms and stemplots are useful to show the shape of a distribution. Both group the data into just a few intervals. Stemplots allow individual data points to be displayed. Below we will illustrate these commands.

The STEM-AND-LEAF Command

A stem-and-leaf plot uses the actual data to create the display. The format for the **STEM-AND-LEAF** command is

```
STEM-AND-LEAF display of C...C
```

The command is illustrated below on the "65 and over" data set. The data is given in Table 1.1 of BPS and is stored in TA01-01.MTW.

```
MTB > stem 'Percent'
```

Character Stem-and-Leaf Display

```
Stem-and-leaf of Percent   N  = 50
Leaf Unit = 0.10

     1     5 2
     1     6
     1     7
     2     8 8
     3     9 9
     5    10 25
    14    11 002244446
   (13)   12 0113445556689
    23    13 0223444578889
    10    14 13445
     5    15 2289
     1    16
     1    17
     1    18 5
```

Selecting **Graph ➤ Stemplot** from the menu can also enter the **STEM-AND-LEAF** command. A dialog box will appear as shown on the following page. When you click on the Variables text box, the valid choices appear in the variable list box. Click on the desired variable and the click Select.

The first column of a stem-and-leaf display is called the depth, the second column holds the stems, and the rest of the display holds the leaves. Each leaf digit represents one observation. In the above stem and leaf display, the first stem is 5 and the first leaf is 2. The corresponding observation is 5.2. The leaf unit at the top of the display tells us where to put the decimal point. In the above example, the Leaf Unit = 0.1, so the decimal point goes before the leaf.

Examining Distributions 15

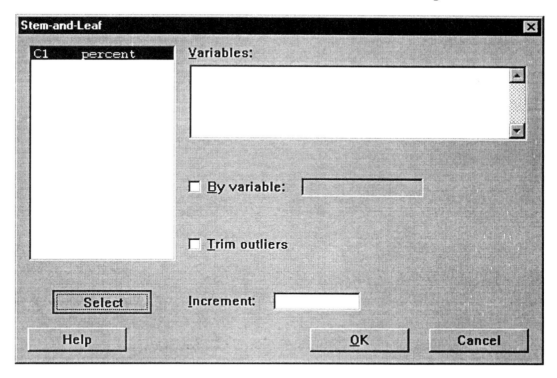

The first column, the depth, has one special line: the line that contains the median. Its value is enclosed in parentheses and indicates how many observations are on just that line. For the remaining lines, the depth gives a cumulative count from the top and bottom. The number five on the sixth line from the top indicates there are five observations on that line and the lines above. They are 5.2, 8.8, 9.9, 10.2, and 10.5.

The **HISTOGRAM** Command

While stem-and-leaf plots are useful for small sets of data, histograms are particularly useful for large data sets. The format for the **HISTOGRAM** command is

HISTOGRAM of C...C

Below we create a histogram for the "65 and over" percents in Table 1.1. To create the histogram, select **Graph ➤ Histogram** from the menu. In the dialog box double click on the variable named percents and click on OK. The session window will display the commands as shown below:

```
MTB > Histogram 'Percent';
SUBC>    MidPoint;
SUBC>    Bar;
SUBC>    ScFrame;
SUBC>    ScAnnotation.
```

In contrast, if you wish to enter the command in the session window, you need only enter:

```
MTB > hist c1
```

Or

```
MTB > hist 'Percent'
```

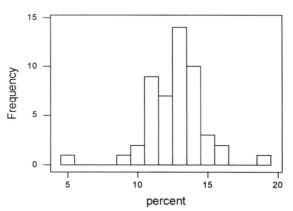

Either way, the histogram will appear as shown. Unlike the stem-and-leaf plot, histograms use the midpoint of a data range for scaling.

Several columns of data can be specified with both of the commands **STEM-AND-LEAF** and **HISTOGRAM**. An individual display will be printed for each column of data listed with the command. We may wish to compare the number of home runs Mark McGwire hit in his first 12 seasons with the number hit by Babe Ruth during his 15 years with the New York Yankees. The data below could be entered into the first two columns of a Minitab worksheet either using the Data window or the Session window.

Ruth   54 59 35 41 46 25 47 60 54 46 49 46 41 34 22
McGwire 49 32 33 39 22 42 9 9 39 52 58 70

Using either the session command HIST 'Ruth' 'McGwire', or the menu command **Graph ➤ Histogram** (with both Ruth and McGwire selected as graph variables), separate histograms are obtained.

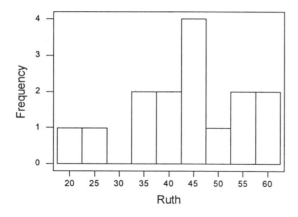
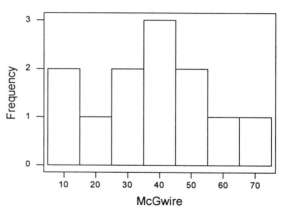

Examining Distributions 17

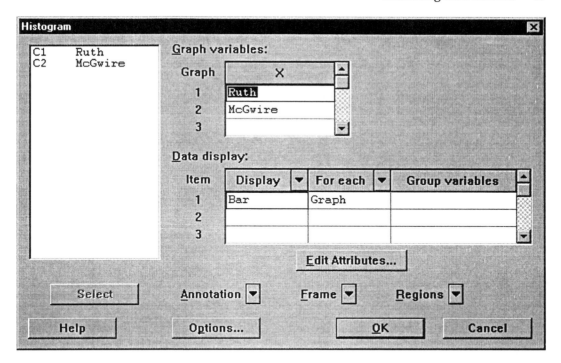

When comparing two or more distributions, the displays should be drawn on axes with the same range. Specifying the midpoints or cutpoints for the histogram intervals can do this. This can be done using the following subcommand:

```
MTB > Histogram 'Ruth' 'McGwire';
SUBC> MidPoint 10 20 30 40 50 60 70.
```

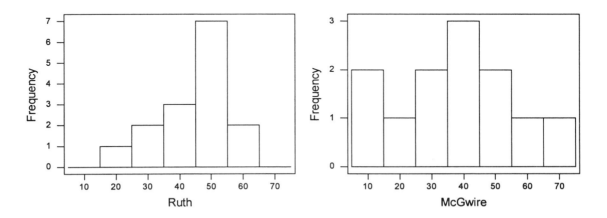

18   *Chapter 1*

Cutpoints or midpoints also can be specified by selecting the Options button on the histogram dialog box. In the Histogram Options dialog box, simply click on MidPoint (or CutPoint) and specify the positions under Definition of Intervals.

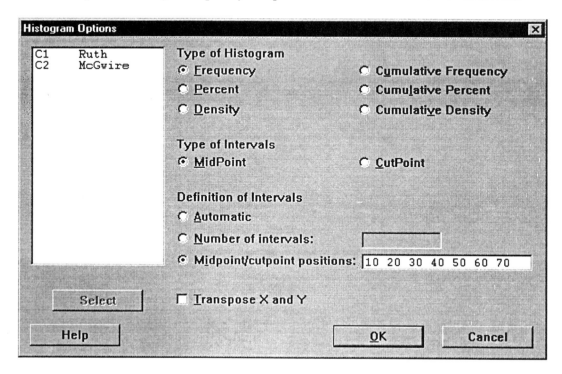

Graphs with exactly the same X and Y axis can be obtained by clicking on Frame on the Histogram dialog box and then selecting Multiple graphs.

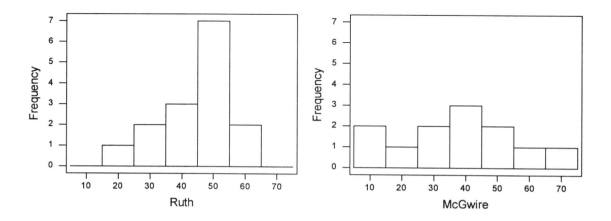

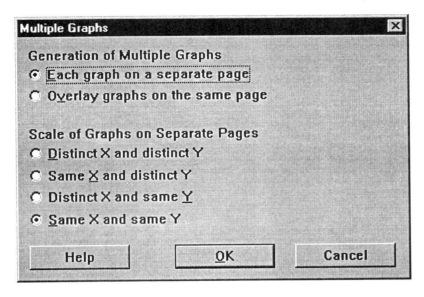

Subcommands

Many commands have subcommands that provide Minitab with additional information. To use a subcommand, put a semicolon at the end of the main command line. This tells Minitab that subcommands will follow. Minitab will prompt the user with the subcommand prompt (SUBC>). If more than one subcommand will be used, start each on a new line and end all except the last with a semicolon. End the last subcommand with a period.

Using **HELP** with a Minitab command name will produce a list of the available subcommands. To obtain more information on a subcommand, click on the subcommand name.

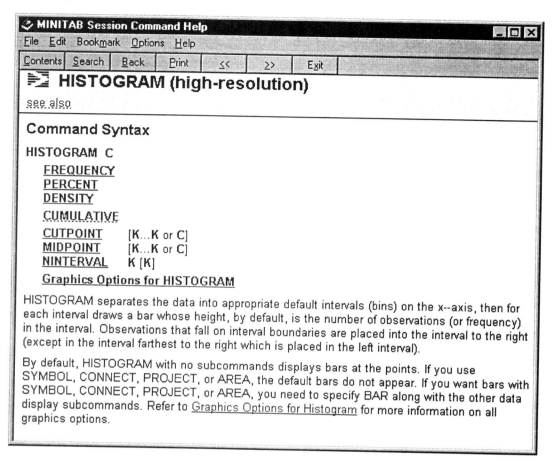

Several columns of data also can be specified with the **STEM-AND-LEAF** command. As with the **HISTOGRAM** command, the scales may come out different for the two plots. To avoid this with stemplots, use the **INCREMENT** subcommand as illustrated below.

```
MTB > stem c1 c2;
SUBC> increment 10.
```

Character Stem-and-Leaf Display

```
Stem-and-leaf of Ruth      N  = 15
Leaf Unit = 1.0

     2     2 25
     4     3 45
    (7)    4 1166679
     4     5 449
     1     6 0
```

```
Stem-and-leaf of McGwire    N  =  12
Leaf Unit = 1.0

     2    0  99
     2    1
     3    2  2
    (4)   3  2399
     5    4  29
     3    5  28
     1    6
     1    7  0
```

The **STACK** Command

Another way to produce stemplots on axes of the same length is with the **BY** subcommand. To use this subcommand, the data must be in a single column with identifying codes in another column. Data can be arranged in this way by using the **STACK** command with the following format.

    STACK (E...E) on ... on (E...E), put in (C...C)

At least three columns must be specified when using the **STACK** command. Minitab will place the contents of the first column on top of the contents of the second column and place the result in the third column. The **SUBSCRIPTS** subcommand can be used to create a column of data that identifies to which column the data originally belonged before the **STACK** command was executed. The **SUBSCRIPTS** subcommand format is

    SUBSCRIPTS put into C

In the following example, we will stack the data for both Babe Ruth and Mark McGwire into C3. The **SUBSCRIPTS** subcommand will place codes into column C4 identifying the baseball player (Ruth = 1, McGwire = 2).

```
MTB > stack 'ruth' on 'mcgwire' put into c3;
SUBC> subscripts c4.
MTB > name c3 'homeruns' c4 'player'
```

Selecting **Manip ➤ Stack/Unstack ➤ Stack Columns** from the menu also stacks the data. Under "Stack the following columns:" enter the columns you want to stack. Under "Store the stacked data in:" enter the appropriate column. Under "Store subscripts in: (Optional)" enter a column where you want to store subscripts. The subscript column will contain 1s in rows corresponding to the first stacked block of columns, 2s in rows corresponding to the second stacked block, and so on.

22   Chapter 1

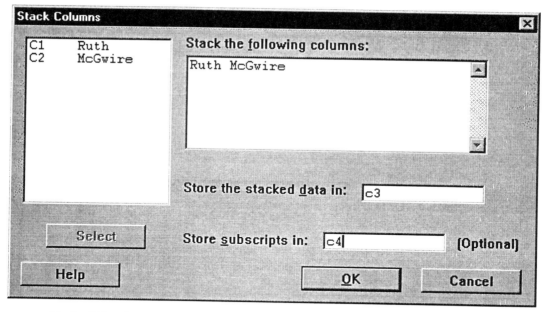

Data Display

Row	homeruns	player
1	54	1
2	59	1
3	35	1
4	41	1
5	46	1
6	25	1
7	47	1
8	60	1
9	54	1
10	46	1
11	49	1
12	46	1
13	41	1
14	34	1
15	22	1
16	49	2
17	32	2
18	33	2
19	39	2
20	22	2
21	42	2
22	9	2
23	9	2
24	39	2
25	52	2
26	58	2
27	70	2

Once the data are arranged in this manner, we will use the **BY** subcommand with the **STEM-AND-LEAF** command to obtain displays using the same axis.

```
MTB > stem c3;
SUBC> by c4.
```

Character Stem-and-Leaf Display

```
Stem-and-leaf of homeruns   player = 1      N = 15
Leaf Unit = 1.0

    1       2 2
    2       2 5
    3       3 4
    4       3 5
    6       4 11
   (5)      4 66679
    4       5 44
    2       5 9
    1       6 0


Stem-and-leaf of homeruns   player = 2      N = 12
Leaf Unit = 1.0

    2       0 99
    2       1
    2       1
    3       2 2
    3       2
    5       3 23
   (2)      3 99
    5       4 2
    4       4 9
    3       5 2
    2       5 8
    1       6
    1       6
    1       7 0
```

The UNSTACK Command

Data such as the home run data in EX01-19.MTW may not be in the format you prefer. In addition to the **STACK** command, you may want to use the **UNSTACK** command. The **UNSTACK** command unstacks, or splits, one column into two or more shorter columns. The command format is

```
UNSTACK (C...C) into (E...E) ... (E...E)
SUBSCRIPTS are in C
```

The subcommand is necessary since the values in the subscript column determine how the source column will be unstacked.

Selecting **Manip ➤ Stack/Unstack ➤ Unstack One Column** from the menu can also unstack the data. Under "Unstack the data in:" enter the column you want to unstack. Under "Store the unstacked data in:" enter the columns where you want to store the unstacked data. The number of columns you list here must equal the number of distinct values in the subscript column below. Under "Using subscripts in:" enter the subscript column. The values in this column determine how the source column will be unstacked.

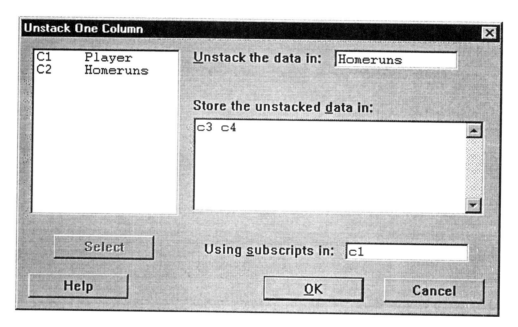

The TSPLOT Command

When data are collected over time, it is a good idea to plot the observations in the order they were collected. The **TSPLOT** command is used to produce time series plots. **TSPLOT** (the TS is for time series) plots the column of data (vertical axis) versus the integers 1, 2, 3, etc. (horizontal axis). The format for the command is listed below:

```
TSPLOT [period = K] of C
```

Babe Ruth played for the New York Yankees for 15 years (1920 to 1934). Below we plotted the number of home runs he hit in order of time. This plot could be produced using the session command

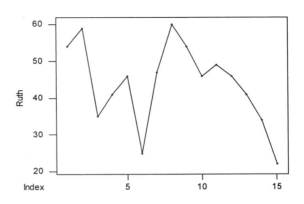

```
MTB > tsplot C1
```

or the menu command **Graph ➤ Time Series Plot**. In the dialog box, the column to be graphed must be selected.

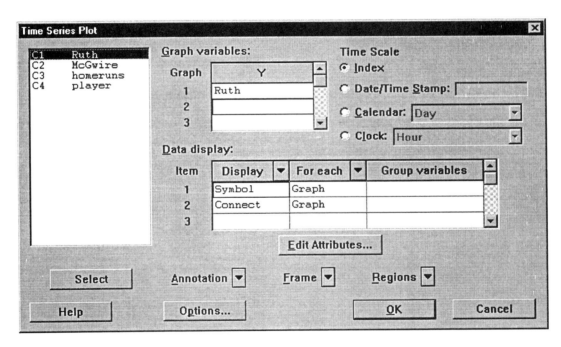

It is important that the observations in time series data occur at regular time intervals when using the `TSPLOT` command. Often in time series data, some observations will be missing. To act as a placeholder for the missing time period, Minitab allows the entry of these missing observations through the use of the asterisk (*).

Instead of indexing the time series by numbers 1 to 15, the years 1920 to 1934 can be used. In the Time Series Plot dialog box, click on the Options button. In the Time Series Plot Options subdialog box enter 1920 as the Start time.

Describing Distributions

Numerical measures are often used to describe distributions. The **DESCRIBE** command summarizes several different measures of both the center and variability of a distribution. The **DESCRIBE** command prints the statistics N, N*, Mean, Median, TrMean, StDev, SE Mean, Minimum, Maximum, Q3, and Q1 for each column specified. The format for the **DESCRIBE** command is

```
DESCRIBE variables in C...C
```

The command can also be used by selecting **Stat ➤ Basic Statistics ➤ Display Descriptive Statistics** from the menu. We illustrate the use of the **DESCRIBE** command using the home run data for Babe Ruth and Mark McGwire. The data for both players are given on page 16.

```
MTB > describe c1 c2
```

Descriptive Statistics

Variable	N	Mean	Median	TrMean	StDev	SE Mean
Ruth	15	43.93	46.00	44.38	11.25	2.90
McGwire	12	37.83	39.00	37.50	18.48	5.34

Variable	Minimum	Maximum	Q1	Q3
Ruth	22.00	60.00	35.00	54.00
McGwire	9.00	70.00	24.50	51.25

N is the number of actual values in the column (missing values are not counted). N* is the number (if any) of missing values. Mean is the average of the values. To find the Median, the data first must be ordered. If N is odd, the Median is the value in the middle. If N is even, the median is the average of the two middle values. The TrMean, or trimmed mean, removes the smallest 5% and the largest 5% of the observations (rounded to the nearest integer) and averages the rest. StDev is the standard deviation computed as

$$\text{StDev}_x = \sqrt{\frac{\sum (x - \bar{x})}{N - 1}}$$

SE Mean is the standard error of the mean. It is calculated as $\text{StDev}/\sqrt{N}$. Q3 is the third quartile and Q1 is the first quartile. Minitab doesn't use exactly the same algorithm to calculate quartiles as BPS, so minor differences in results will sometimes occur.

The **BY** subcommand can also be used with the **DESCRIBE** command as illustrated below.

```
MTB > describe c3;
SUBC> by c4.
```

Descriptive Statistics

Variable	player	N	Mean	Median	TrMean	StDev
homeruns	1	15	43.93	46.00	44.38	11.25
	2	12	37.83	39.00	37.50	18.48

Variable	player	SE Mean	Minimum	Maximum	Q1	Q3
homeruns	1	2.90	22.00	60.00	35.00	54.00
	2	5.34	9.00	70.00	24.50	51.25

The commands **N**, **NMISS**, **MEAN**, **MEDIAN**, **STDEV**, **MINIMUM**, and **MAXIMUM** can also be used to obtain these measures for a column of data. The commands can be used by selecting **Calc ➤ Column Statistics** from the menu or by using the following command formats.

```
COUNT the number of values in C [put into K]
N count the nonmissing values in C [put into K]
NMISS (number of missing values in) C [put into K]
SUM of the values in C [put into K]
MEAN of the values in C [put into K]
MEDIAN of the values in C [put into K]
MINIMUM of the values in C [put into K]
MAXIMUM of the values in C [put into K]
```

In each command, the answer is printed and may also be stored. The command

```
MTB > mean c1 k1
```

Column Mean

```
    Mean of Ruth = 43.933
```

will put the value 43.933 into the constant K1.

Rowwise Statistics:

The statistics we have just described are also available for rows. These commands compute summaries across rows rather than down columns. The answers are always stored in a column. The commands can be used by selecting **Calc ➤ Row Statistics** from the menu or by using the following command formats.

```
RCOUNT  of  E...E  put  into  C
RN      of  E...E  put  into  C
RNMISS  of  E...E  put  into  C
RSUM    of  E...E  put  into  C
RMEAN   of  E...E  put  into  C
RSTDEV  of  E...E  put  into  C
RMEDIAN of  E...E  put  into  C
RMINIMUM of E...E  put  into  C
RMAXIMUM of E...E  put  into  C
```

For the commands **RMEAN, RMEDIAN, RSTDEV, RMAXIMUM,** and **RMINIMUM,** as well as for **MEAN, MEDIAN, STDEV, MAXIMUM,** and **MINIMUM,** missing observations are omitted from the calculations.

The BOXPLOT Command

The five-number summary consisting of the median, quartiles, and minimum and maximum values provides a quick overall description of a distribution. Boxplots based on the five-number summary display the main features of a column of data. Boxplots can be obtained with the following format or by selecting **Graph ➤ Boxplot** from the menu.

```
BOXPLOT  C...C
```

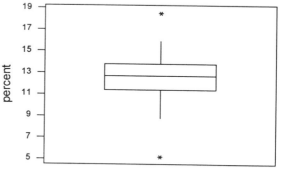

A boxplot graphically displays the main features of data from a single variable. The **BOXPLOT** command is illustrated below on the "65 and over" data set. The data are given in Table 1.1 of BPS and in TA01-01.MTW.

The boxplot consists of a box, whiskers, and outliers. Minitab draws a line across the box at the median. By default, the bottom of the box is at the first quartile (Q1) and the top is at the third quartile (Q3). The whiskers are the lines that extend from the top and bottom of the box to the adjacent values, the lowest and highest observations still inside the region defined by the lower limit $Q1 - 1.5(Q3 - Q1)$ and the upper limit $Q3 + 1.5(Q3 - Q1)$. Outliers are points outside the lower and upper limits, plotted with asterisks (*).

The simplest form, **BOXPLOT C**, generates just one boxplot. In the form **BOXPLOT C*C**, the second variable is a category (or grouping) variable. This form produces side-by-side boxplots for the group or categories in the second column.

```
MTB > boxp 'homeruns'*'player'
```

produces the graph as shown. Here player 1 is Babe Ruth and player 2 is Mark McGwire. The categories column may be numeric or text. If "Ruth" and "McGwire" were used in the category column, the players names would replace the 1 and 2 on the graph.

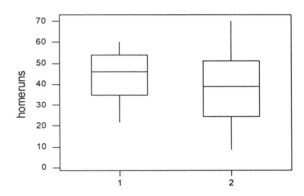

Selecting **Graph ➤ Boxplot** from the menu and filling in the dialog box can also produce boxplots. For a boxplot of one variable, select a column for the box below Y. For side-by-side boxplots, select measurements (homeruns) into Y and a categorical variable (player) into X.

Normal Calculations

Minitab can be used to perform normal distribution calculations. If data in a column are normally distributed, then the **CENTER** command can be used to obtain the standardized values, that is, those with mean equal to zero and standard deviation equal to one. The format for the **CENTER** command is

```
CENTER the data in C...C put into C...C
```

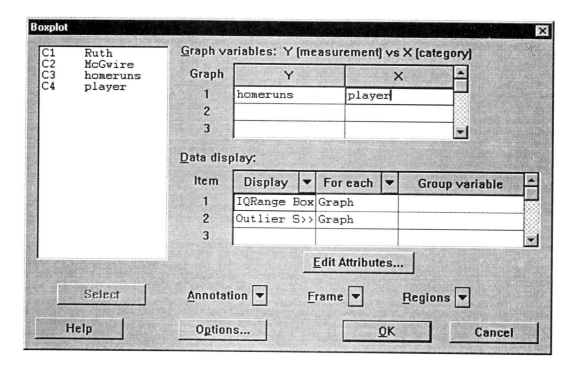

The command can also be used by selecting **Calc ➤ Standardize** from the menu. This command transforms each input column by subtracting its mean and dividing by its standard deviation as illustrated below. In C1, we have stored the exam scores for a class of 40 students. Using the **CENTER** command, the standardized values $z = (x - \bar{x})/s$ can be stored in C2 as illustrated below.

```
MTB > print c1
```

Data Display

```
C1
   58    98    57    70    60    62    74    75    81    99
   49    75    71    86    73    90    82    74    63    61
   55    80    52   100    75    58    63    66    68    55
  100    91   100    66    84    36    90    75    89    78
   84    84    87    83    79    65    43    65    75    78
```

```
MTB > center c1 c2
MTB > print c2
```

Data Display

```
C2
-1.02907    1.59946   -1.09478   -0.24051   -0.89764   -0.76622
 0.02234    0.08806    0.48234    1.66517   -1.62049    0.08806
-0.17480    0.81090   -0.04337    1.07376    0.54805    0.02234
-0.70050   -0.83193   -1.22621    0.41662   -1.42335    1.73089
-1.02907   -0.70050   -0.50336    0.37194   -1.22621    1.79660
 1.13947    1.73089   -0.50336    0.67948   -2.47476    1.07376
 0.08806    1.00804    0.28520    0.67948    0.67948    0.87662
 0.61376    0.35091   -0.56908   -2.01477   -0.56908    0.08806
 0.28520
```

The standardized test scores will tell how far above or below the mean an individual scored. The measure is in units of standard deviations. The first score, 58, is just slightly more than one standard deviation ($z = -1.02907$) below the mean. The second score, 98, is nearly 1.6 standard deviations ($z = 1.59946$) above the mean.

The LET Command

Recall that in the introductory chapter, the **LET** command was used to correct individual data point entries. In addition to that function, the **LET** command can be used to do arithmetic using the following operations.

+	Addition
−	Subtraction
*	Multiplication
/	Division
**	Exponentiation

Additional Minitab functions such as **MEAN**, **STDEV**, **ROUND**, **SQRT**, **LOGTEN**, **LOGE**, **EXPONENTIATE**, **N**, **MAXIMUM**, and **MINIMUM** also can be used as part of the expression on the right-hand side of the **LET** command.

The **LET** command can be used to find the standardized value of an observation. In example 1.14 of BPS, we found the standardized height of a 68-inch-tall woman. The heights of young women are approximately normal with $\mu = 64.5$ and $\sigma = 2.5$ inches. The standardized height is

$$z = \frac{68 - 64.5}{2.5} = 1.4$$

or 1.4 standard deviations above the mean. The calculation could be done in Minitab as

```
MTB > let k1 = (68-64.5)/2.5
MTB > print k1
K1        1.40000
```

The calculator also lets you perform mathematical functions. The results of the calculation can be stored in a column or constant. To use the calculator, choose **Calc ➤ Calculator** from the menu. Under Store result in variable, enter a new or existing column or constant. Under Expression, select variables and functions from their respective lists, and click calculator buttons for numbers and arithmetic functions. You can also type the expressions.

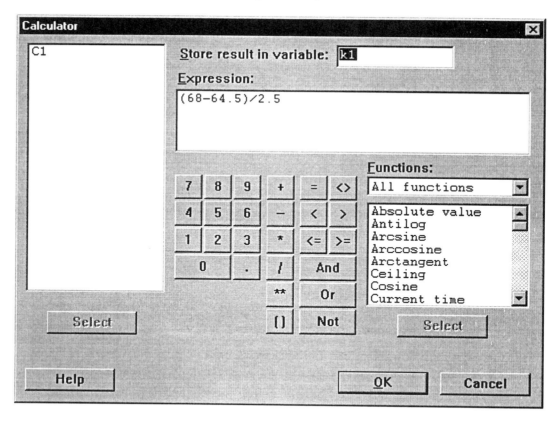

The CDF Command

The CDF command calculates the area under the standard normal curve to the left of a specified value of z. The command format is

    CDF for values in E...E [put results in E...E]

If a subcommand is not used with the CDF command, a standard normal distribution with $\mu = 0$ and $\sigma = 1$ is assumed. That is, if X has the $N(\mu, \sigma)$ distribution, then the standardized variable $z = (X - \mu)/\sigma$ has the standard normal distribution $N(0,1)$. Relative frequencies for the event $Z < z$ can be calculated using the CDF command instead of Table A in BPS. To find the proportion of all young women who are less than 68 inches tall, we have stored the standardized value $z = 1.4$ in K1. We now find the area to the left of 1.40.

    MTB > cdf k1 k2
    MTB > print k1 k2
    K1   1.40000
    K2   0.919243

The same normal distribution calculation could also be done without standardizing the observation if a subcommand is used with the CDF command to specify that a $N(64.5, 2,5)$ distribution is to be used. The subcommand format is given below.

    NORMAL [mu = K [sigma = K]]

The normal distribution has parameters mu and sigma. If you omit the arguments, $\mu = 0$ and $\sigma = 1$ are used. The CDF command with the NORMAL subcommand is illustrated below.

    MTB > cdf 68;
    SUBC> norm 64.5 2.5.
        68.0000   0.9192

The INVCDF Command

The INVCDF command is useful to help find a value given a proportion. If we want to find the value with a given proportion of the observations below it, we use Table A backward. The INVCDF command does this for us by finding the standardized value of z for a specified proportion. The command format is

    INVCDF for values in E [put into E]

In Example 1.18 of BPS, we want to know how high a student must score in order to place in the top 10% of all students taking the SAT for verbal ability. The scores of high school seniors follow the $N(505,110)$ distribution. We need the z value with 90 percent of the observations below it, so we use the **INVCDF** command as follows.

```
MTB > invc .9 k1
MTB > print k1
```

Data Display

```
K1    1.28155
```

We now can use the **LET** command to solve the equation

$$\frac{x-505}{110} = 1.2816.$$

```
MTB > let k2 = k1*110+505
MTB > print k2
```

Data Display

```
K2    645.971
```

A student must score at least 646 to have 90% of the students scoring below him and 10% above. Alternatively, we can use the **INVCDF** with the **NORMAL** subcommand to specify the $N(505, 110)$ distribution. This gives the same result as above.

```
MTB > invcdf .9;
SUBC> norm 505, 110.
```

Inverse Cumulative Distribution Function

```
Normal with mean = 505.000 and standard deviation = 110.000

P( X <= x)         x
   0.9000      645.9707
```

You can use Minitab to do normal probability calculations by selecting **Calc ➤ Probability Distributions ➤ Normal** from the menu. Choose either Cumulative probability or Inverse cumulative probability from the dialog box and specify either constants or columns for your calculations.

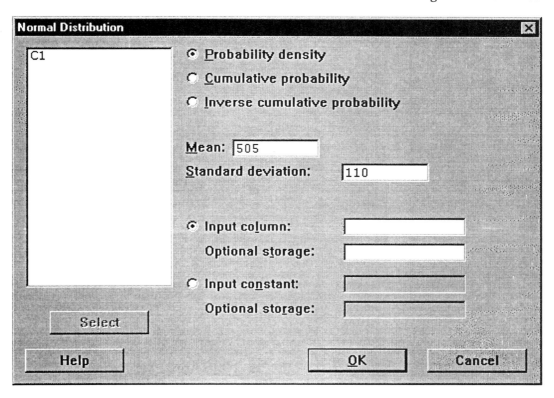

EXERCISES

1.5 Environmental Protection Agency regulations require automakers to give the city and highway gas mileages for each model of car. Table 1.2 in BPS and TA01-02.MTW give the highway mileages (miles per gallon) for 26 midsize 1998 car models.
   (a) Make a **HISTOGRAM** of the highway mileages of these cars.
   (b) Describe the main features (shape, center, spread, outliers) of the distribution of highway mileage.
   (c) The government imposes a "gas guzzler" tax on cars with low gas mileage. Which of these cars do you think are subject to the gas guzzler tax?

1.9 The Survey of Study Habits and Attitudes (SSHA) is a psychological test that evaluates college students' motivation, study habits, and attitudes toward school. A private college gives the SSHA to 18 of its incoming first-year female students. Their scores are given in EX01-09.MTW and below.

36 Chapter 1

154	109	137	115	152	140	154	178	101
103	126	126	137	165	165	129	200	148

(a) Make a **STEM-AND-LEAF** display of these data. The overall shape of the distribution is irregular, as often happens when only a few observations are available. Are there any outliers?

(b) Use the **DESCRIBE** command to find the mean score and the median score for this group. Explain the relationship between these two measures in terms of the main features of the distribution of scores.

(c) The stemplot suggests that the score 200 is an outlier. Use the **DESCRIBE** command to find the mean and median for the 17 observations that remain when you drop the outlier. Briefly describe how the outlier changes the mean and median.

1.10 Many people invest in "money market funds." These are mutual funds that attempt to maintain a constant price of $1 per share while paying monthly interest. Here and in EX01-10.MTW are the average annual interest rates (in percent) paid by all taxable money market funds since 1973, the first full year in which such funds were available.

Year	Rate	Year	Rate	Year	Rate	Year	Rate
1973	7.60	1979	10.92	1985	7.77	1991	5.70
1974	10.79	1980	12.88	1986	6.30	1992	3.31
1975	6.39	1981	17.16	1987	6.17	1993	2.62
1976	5.11	1982	12.55	1988	7.09	1994	3.65
1977	4.92	1983	8.69	1989	8.85	1995	5.37
1978	7.25	1984	10.21	1990	7.81	1996	4.80

(a) Make a **TIMEPLOT** of the interest paid by money market funds for these years.

(b) Interest rates, like many economic variables, show clear but irregular up-and-down movements called cycles. In which years did the interest rate cycle reach temporary peaks?

(c) A time plot may show a consistent trend underneath cycles. When did interest rates reach their overall peak during these years? Has there been a general trend downward since that year?

Table 1.6 in BPS and TA01-06.MTW present data about the individual states that relate to education. Study of a data set with many variables begins by examining each variable by itself. Exercises 1.24 to 1.26 concern the data in TA01-06.MTW.

1.24 Make a **STEM-AND-LEAF** display of the population of the states. Briefly describe the shape, center, and spread of the distribution of population. Explain why the shape of the distribution is not surprising. Are there any states that you consider outliers?

1.25 Make a **STEM-AND-LEAF** display of the distribution of the percent of high school seniors who take the SAT in the various states. Briefly describe the overall shape of the distribution. Find the midpoint of the data and mark this value on your stemplot. Explain why describing the center is not very useful for a distribution with this shape.

1.26 Make a **HISTOGRAM** to display the distribution of average teachers' salaries for the states. Is there a clear overall pattern? Are there any outliers or other notable deviations from the pattern?

1.28 A study in Switzerland examined the number of caesarean sections (surgical deliveries of babies) performed by doctors in a year. The data for 15 male doctors is given in EX01-28.MTW and below.

27 50 33 25 86 25 85 31 37 44 20 36 59 34 28

(a) Make a **STEM-AND-LEAF** display of the data. Note the two high outliers.

(b) Use the **DESCRIBE** command to find the mean and median number of operations. How do the outliers explain the difference between your two results?

(c) Find the mean and median number of operations without the two outliers. How does comparing your results in (b) and (c) illustrate the resistance of the median and the lack of resistance of the mean?

1.31 Exercise 1.28 and EX01-28.MTW give the number of caesarean sections performed by 15 male doctors in Switzerland. The study also looked at 10 female doctors. The numbers of caesareans performed by these doctors (arranged in order) is given in EX01-31.MTW and below.

5 7 10 14 18 19 25 29 31 33

(a) Use the **DESCRIBE** command to find the five-number summary for each group.

(b) Copy the data from EX01-28.MTW onto your worksheet. Make side-by-side **BOXPLOTS** to compare the number of operations performed by female and male doctors. What do you conclude?

1.32 How old are presidents at their inaugurations? Was Bill Clinton, at age 46, unusually young? Table 1.7 in BPS and TA01-07.MTW give the ages of all U.S. presidents when they took office.

57	61	57	57	58	57	61	54	68	51	49	64	50	48
65	52	56	46	54	49	51	47	55	55	54	42	51	56
55	51	54	51	60	61	43	55	56	61	52	69	64	46

(a) Make a **STEM-AND-LEAF** display of the distribution of ages. From the shape of the distribution, do you expect the median to be much less than the mean, about the same as the mean, or much greater than the mean?

(b) Use the **DESCRIBE** command to find the mean and the five-number summary. Verify your expectation about the median.

(c) What is the range of the middle half of the ages of new presidents? Was Bill Clinton in the youngest 25%?

1.38 Here and in EX01-38.MTW are the percents of the popular vote won by the successful candidate in each of the presidential elections from 1948 to 1996:

Year	1948	1952	1956	1960	1964	1968	1972	1976	1980	1984	1988	1992	1996
%	49.6	55.1	57.4	49.7	61.1	43.4	60.7	50.1	50.7	58.8	53/9	43.2	49.2

(a) Make a **STEM-AND-LEAF** display of the winners' percents.

(b) What is the median percent of the vote won by the successful candidate in presidential elections? Use the **DESCRIBE** command.

(c) Call an election a landslide if the winner's percent falls at or above the third quartile. Find the third quartile. Which elections were landslides?

1.39 Some people worry about how many calories they consume. *Consumer Reports* magazine, in a story on hot dogs, measured the calories in 20 brands of beef hot dogs, 17 brands of meat hot dogs, and 17 brands of poultry hot dogs. The data are give below and in EX01-39.MTW.

beef	186	181	176	149	184	190	158	139	175	148
	152	111	141	153	190	157	131	149	135	132
meat	173	191	182	190	172	147	146	139	175	136
	179	153	107	195	135	140	138			
poultry	129	132	102	106	94	102	87	99	107	113
	135	142	86	143	152	146	144			

(a) Use the **DESCRIBE** command to summarize the distributions of calories for the three types of hot dogs.

(b) Make **BOXPLOTS** using the same scales for the three types of hot dogs.

(c) Write a brief comparison of the distributions. Will eating poultry hot dogs usually lower your calorie consumption compared with eating beef or meat hot dogs?

1.41 In 1798 the English scientist Henry Cavendish measured the density of the earth with great care. It is common practice to repeat careful measurements several times and use the mean as the final result. Cavendish repeated his work 29 times. His results (the data give the density of the earth as a multiple of the density of water) are given below and in EX01-41.MTW.

5.50	5.61	4.88	5.07	5.26	5.55	5.36	5.29	5.58	5.65
5.57	5.53	5.62	5.29	5.44	5.34	5.79	5.10	5.27	5.39
5.42	5.47	5.63	5.34	5.46	5.30	5.75	5.68	5.85	

(a) Present these measurements with a **STEM-AND-LEAF**. Does the shape of the distribution allow the use of the mean and standard deviation to describe it?

(b) Use the **DESCRIBE** command to find $\bar{x}$ and s. What is your estimate of the density of the earth based on these measurements?

1.45 The Baltimore Orioles had the highest team payroll in major league baseball in 1998. Here and in EX01-45.MTW are the salaries of the Orioles' players in thousands of dollars. The number, 6495 stands for Mike Mussina's salary of $6,495,000.

6495	6486	6300	6269	5442	5391	3600	3600	3583
3089	2850	2500	1950	1663	1367	1333	1150	900
856	800	800	665	650	450	450	170	170

Describe this salary distribution both with a graph and with an appropriate numerical summary. Then write a brief description of the important features of the distribution.

(a) What percent of men are at least 6 feet (72 inches) tall?
(b) What percent of men are between 5 feet (60 inches) and 6 feet tall?
(c) How tall must a man be to be in the tallest 10% of all adult men?

1.57 Use the **CDF** command to find the proportion of observations from a standard normal distribution that satisfies each of the following statements. In each case, sketch a standard normal curve and shade the area under the curve that is the answer to the question.
(a) $z < 2.85$
(b) $z > 2.85$
(c) $z > -1.66$
(d) $-1.66 < z < 2.85$

1.59 Use the **INVCDF** command to find the value z of a standard normal variable that satisfies each of the following conditions. In each case, sketch a standard normal curve with your value of z marked on the axis.
(a) The point z with 25% of the observations falling below it.
(b) The point z with 40% of the observations falling above it.

1.60 Scores on the Wechsler Adult Intelligence Scale (a standard "IQ test") for the 20 to 34 age group are approximately normally distributed with mean 110 and standard deviation 25. Use Minitab to answer the following questions.
(a) What percent of people age 20 to 34 have IQ scores above 100?
(b) What IQ scores fall in the lowest 25% of the distribution?
(c) How high an IQ score is needed to be in the highest 5%?

1.75 Professor Moore, who lives a few miles outside a college town, records the time he takes to drive to the college each morning. Below and in EX01-75.MTW are the times (in minutes) for 42 consecutive weekdays, with the dates in order along the rows.

8.25 7.83 8.30 8.42 8.50 8.67 8.17 9.00 9.00 8.17 7.92
9.00 8.50 9.00 7.75 7.92 8.00 8.08 8.42 8.75 8.08 9.75
8.33 7.83 7.92 8.58 7.83 8.42 7.75 7.42 6.75 7.42 8.50
8.6 10.17 8.75 8.58 8.67 9.17 9.08 8.83 8.67

(a) Make a **STEM-AND-LEAF** display of these drive times. Is the distribution roughly symmetric, clearly skewed, or neither? Are there any clear outliers?

(b) Make a **TSPLOT** of the drive times. The plot shows no clear trend, but it does show one unusually low drive time and two unusually high drive times. Circle these observations on your plot.

(c) All three unusual observations can be explained. The low time is the day after Thanksgiving (no traffic on campus). The two high times reflect delays due to an accident and icy roads. **DELETE** these three observations, and use the **DESCRIBE** command to find the mean, $\bar{x}$ and standard deviation, s, of the remaining 39 drive times.

(d) Check how closely the data follow the 68–95–99.7 rule. Count the number of observations that fall between $\bar{x} - s$ and $\bar{x} + s$, between $\bar{x} - 2s$ and $\bar{x} + 2s$, and between $\bar{x} - 3s$ and $\bar{x} + 3s$. Find the percent of observations in each of these intervals and compare with the 68–95–99.7 rule. As real data go, these data are reasonably close to having a normal distribution.

1.77 Corn is an important animal food. Normal corn lacks certain amino acids, which are building blocks for protein. Plant scientists have developed new corn varieties that have more of these amino acids. To test a new corn as an animal food, a group of 20 one-day-old male chicks was fed a ration containing the new corn. A control group of another 20 chicks was fed a ration that was identical except that it contained normal corn. The weight gains (in grams) after 21 days are below and in EX01-77.MTW.

Normal corn				New corn			
380	321	366	356	361	447	401	375
283	349	402	462	434	403	393	426
356	410	329	399	406	318	467	407
350	384	316	272	427	420	477	392
345	455	360	431	430	339	410	326

(a) Use the **DESCRIBE** command for the weight gains of the two groups of chicks. Then make side-by-side **BOXPLOTS** to compare the two distributions. What do the data show about the effect of the new corn?

(b) The researchers actually reported means and standard deviations for the two groups of chicks. What are they? How much larger is the mean weight gain of chicks fed the new corn?

1.78 Joe DiMaggio played center field for the Yankees for 13 years. Mickey Mantle, who played for 18 years, succeeded him. The number of home runs hit each year by DiMaggio and Mantle are given below and in EX01-78.MTW.

DiMaggio:
29 46 32 30 32 30 21 25 20 39 14 32 12

Mantle:
13 23 21 27 37 52 34 42 31 40 54 30 15 35 19 23 22 18

Make side-by-side **BOXPLOTS** of the home run distributions. What does your comparison show about DiMaggio and Mantle as home run hitters?

1.79 EX01-79.MTW gives the highway fuel consumption (in miles per gallon) for 26 midsize 1998 cars and 19 four-wheel drive 1998 sport utility vehicles.

(a) Give a graphical and numerical description of highway fuel consumption for sports utility vehicles. What are the main features of the distribution?

(b) Make side-by-side **BOXPLOTS** to compare the highway fuel consumption of midsize cars and sports utility vehicles. What are the most important differences between the two distributions?

1.80 The data below are the survival times of 72 guinea pigs after they were injected with tubercle bacilli in a medical experiment. These data are stored in TA01-09.MTW. Survival times, whether of machines under stress or cancer patients after treatment, usually have distributions that are skewed to the right.

43	45	53	56	56	57	58	66	67
73	74	79	80	80	81	81	81	82
83	83	84	88	89	91	91	92	92
97	99	99	100	100	101	102	102	102
103	104	107	108	109	113	114	118	121
123	126	128	137	138	139	144	145	147
156	162	174	178	179	184	191	198	211
214	243	249	329	380	403	511	522	598

(a) Graph the distribution using a display of your choice and describe its main features. Does it show the expected right skew?

(b) Use the **DESCRIBE** command to find the mean and the median. Explain how the relationship between the mean and the median reflects the skewness of the data.

1.81 The rate of return on a stock is its change in price plus any dividends paid, usually measured in percent of the starting value. We have data on the monthly rate of return for the stock of Wal-Mart stores for the years 1973 to 1991, the first 19 years Wal-Mart was listed on the New York Stock Exchange. There are 228 observations stored in TA01-08.MTW.

(a) Make a **STEM-AND-LEAF** display and use the **DESCRIBE** command to find the five-number summary for monthly returns on Wal-Mart stock.

(b) Describe in words the main features of the distribution.

(c) If you had $1000 worth of Wal-Mart stock at the beginning of the best month during these 19 years, how much would your stock be worth at the end of the month? If you had $1000 worth of stock at the beginning of the worst month, how much would your stock be worth at the end of the month?

1.86 Of the 50 species of oaks in the United States, 28 grow on the Atlantic coast and 11 grow in California. We are interested in the distribution of acorn sizes among oak species. Here and in EX01-86.MTW are data on the volume of acorns (in cubic centimeters) for these 39 oak species.

Atlantic							California		
1.4	3.4	9.1	1.6	10.5	2.5	0.9	4.1	5.9	17.1
6.8	1.8	0.3	0.9	0.8	2.0	1.1	1.6	2.6	0.4
0.6	1.8	4.8	1.1	3.0	1.1	1.1	2.0	6.0	7.1
3.6	8.1	3.6	1.8	0.4	1.1	1.2	5.5	1.0	

(a) Make a **HISTOGRAM** of the 39 acorn sizes. Describe the distribution and include an appropriate numerical description.

(b) Compare the Atlantic and California region distributions with a graph and numerical summaries. You will need to **UNSTACK** the data to make the histograms. What do you find?

Chapter 2
Examining Relationships

Commands to be covered in this chapter:

```
PLOT C * C
CORRELATION between C...C
REGRESS C on K predictors C...C
TABLE the data classified by C...C
```

Scatterplots

Often we are interested in illustrating the relationships between two variables, such as the relationship between height and weight, between smoking and lung cancer, or between advertising expenditures and sales. For illustration, we will consider the relationship between the percent of students in a state who take the SAT exam and the state average SAT mathematics score, using data from Table 1.6 in BPS. The data are also stored in TA01-06.MTW.

The PLOT Command

If both variables are quantitative, the most useful display of their relationship is the scatterplot. Scatterplots can be produced using the **PLOT** command. The command format is

```
PLOT C * C
```

The first column is the vertical axis; the second is the horizontal axis. So, to obtain a scatterplot with "percent taking" as the explanatory or *x* variable and "average score" as the *y* variable, we can type the following command.

```
MTB > plot 'SAT math'*'Percent taking'
```

Alternatively, a scatterplot can be obtained by selecting **Graph ➤ Plot** from the menu. In the dialog box, SAT math and Percent taking are selected as the *y* and *x* variables, respectively.

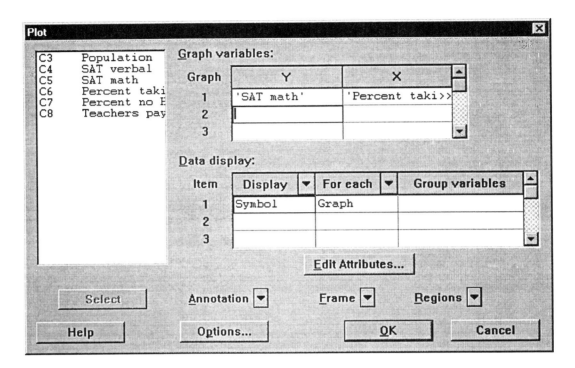

Using either the session command or the menu command, the scatterplot will appear as shown. There are two distinct clusters with a gap between them.

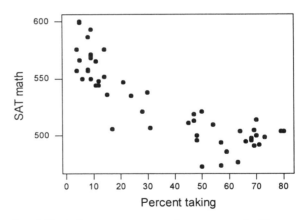

We can add information about a third categorical variable to a scatterplot by using different symbols for different points. This can easily be done using either the menu command or the session command. With the session command, adding the **SYMBOL** subcommand allows different categories to be plotted with different plotting symbols. The format for the subcommand is

    SYMBOL C

where C is a column that identifies the categories. The subcommand is illustrated below the points in different regions plotted with different symbols.

```
MTB > Plot 'SAT math'*'Percent taking';
SUBC>    Symbol  'Region'.
```

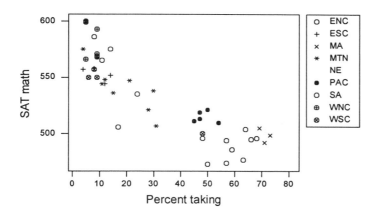

The same labeled scatterplot could be obtained by selecting **Graph ➤ Plot** from the menu and then changing For Each to Group and entering the Group variable and then clicking OK.

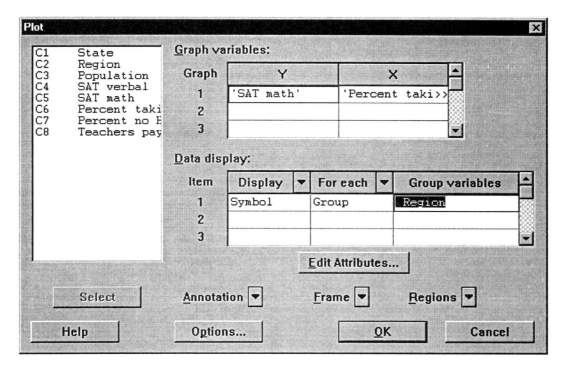

The CORRELATION Command

The **CORRELATION** command computes the correlation coefficient between two columns of data. If more than two columns are specified in the following format, Minitab will print a table giving the correlation coefficients between all pairs of columns.

```
CORRELATION between C...C
```

The command can also be executed by selecting **Stat ➤ Basic Statistics ➤ Correlation** and selecting the desired variables.

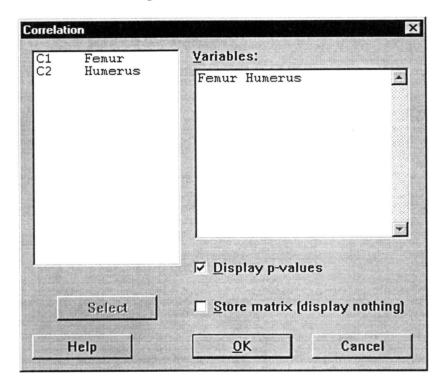

Below we illustrate the use of the **CORRELATION** command with data from Exercise 2.17 of BPS. The data represent the lengths in millimeters for a leg bone (femur) and a bone in the forearm (humerus) for five fossil specimens of *Archaeopteryx*. The data are given below and in EX02-17.MTW.

```
MTB > print c1 c2
```

Data Display

```
Row    Femur    Humerus

 1       38       41
 2       56       63
 3       59       70
 4       64       72
 5       74       84
```

```
MTB > corr c1 c2
```

Correlations (Pearson)

```
Correlation of Femur and Humerus = 0.994, P-Value = 0.001
```

The REGRESS Command

The scatterplot below shows that there is a strong linear relationship between the average outside temperature (measured by heating degree-days) in a month and the average amount of natural gas that the Sanchez household uses per day during the month. This relationship is discussed in Example 2.8 of BPS. The data are given in TA02-01.MTW.

The Minitab REGRESS command will calculate a least-squares line of the form $y = a + bt$ from data. The format for the REGRESS command is

```
REGRESS C on K predictors C...C
```

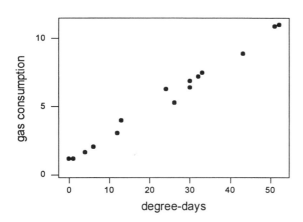

The first column is the y variable, or the variable that we want to predict (gas consumption). The second column is the predictor (degree-days), or x variable. Letting $k = 1$ indicates that only one predictor variable is being used. Below we illustrate the use of the REGRESS command.

```
MTB > regr c2 1 c1
```

Regression Analysis

```
The regression equation is
gas consumption = 1.09 + 0.189 degree-days

Predictor        Coef       StDev         T         P
Constant       1.0892      0.1389      7.84     0.000
degree-d     0.188999    0.004934     38.31     0.000

S = 0.3389      R-Sq = 99.1%     R-Sq(adj) = 99.0%

Analysis of Variance

Source            DF         SS         MS         F         P
Regression         1     168.58     168.58   1467.55     0.000
Residual Error    14       1.61       0.11
Total             15     170.19
```

Identical output can be obtained by selecting **Stat ➤ Regression ➤ Regression** from the menu and entering Degree-days for the Predictor and Gas consumption for the Response in the dialog box.

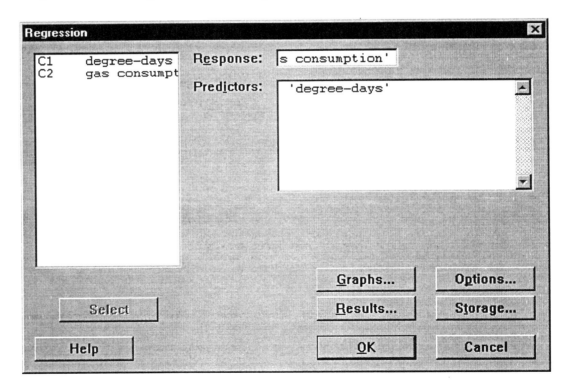

The output indicates that the least-squares regression equation is gas consumption = 1.09 + 0.189 degree-days. Below this equation, the output indicates that $a = 1.0892$ and $b = 0.188999$ are more accurate values of the regression parameters. These values are located under the column labeled Coef. The square of the correlation coefficient, r^2, appears in the output from the **REGRESS** command. It is listed below as a percentage (R-sq = 99.1%). Other useful information is provided and will be discussed in Chapter 10.

Fitted line plots can be obtained by selecting **Stat ➤ Regression ➤ Fitted line plot** from the menu and entering the appropriate predictor and response variable.

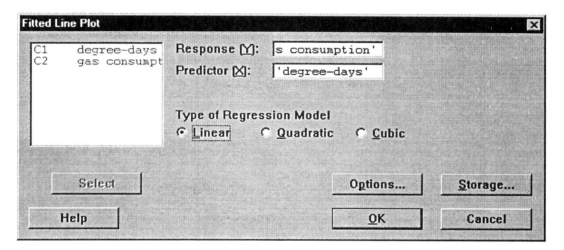

The fitted values (gas consumption = 1.09 + 0.189 degree-days) and the residuals (residual = observation − fit) for the above regression can be obtained using the **LET** command and the least-squares regression equation. C2 contains the data for the response variable, "gas consumption"; C1 contains the data for the explanatory variable, "degree-days." The residuals from the regression can be calculated by first finding the fitted values using the regression equation, then subtracting those values from the actual observations.

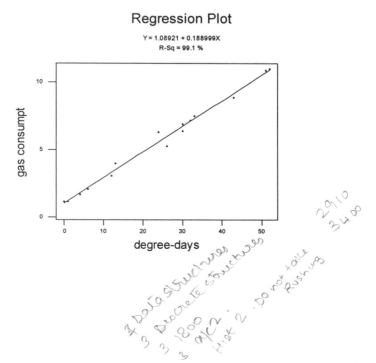

52   Chapter 2

```
MTB > let c3 = 1.0892 + .18899*c1
MTB > let c4 = c2-c3
MTB > name c3 'fits' c4 'resids'
MTB > print c3
```

Data Display

```
fits
    5.6250    10.7277    9.2158    7.3259    6.0029    3.5461    1.8452
    1.0892     1.0892    1.2782    2.2231    3.3571    6.7589    7.1369
   10.9167     6.7589
```

```
MTB > print c4
```

Data Display

```
resids
    0.675040    0.172310   -0.315770    0.174130   -0.702940    0.453930
   -0.145160    0.110800    0.110800   -0.078190   -0.123140   -0.257080
   -0.358900    0.063120    0.083320    0.141100
```

The fitted values and residuals can also be obtained using the **FITS** and **RESIDUALS** subcommands. The formats are:

FITS put into C
RESIDUALS put into C

Alternatively, after selecting **Stat ➤ Regression ➤ Regression**, click on the Storage button and select Fits and Residuals.

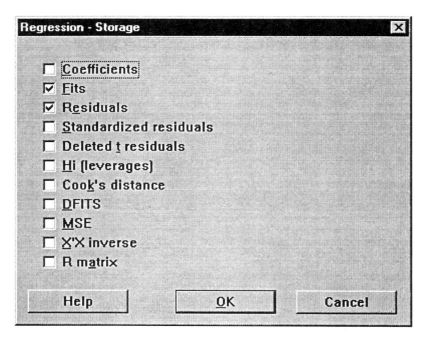

A residual plot is obtained by plotting the residuals versus degree-days, the explanatory variable with the **PLOT** command or by selecting **Graph ➤ Plot** from the menu.

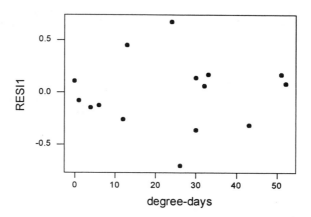

In Example 2.8 in BPS, the Sanchez household wants to use the regression relationship to predict their natural gas consumption. If a month averages 20 degree-days per day (that's 45° F), how much gas will they use? This question can be answered with the **PREDICT** subcommand with the **REGRESS** command. The subcommand format is

PREDICT for E...E

The explanatory variable may be specified as a constant such as 20 or K3.

```
MTB > regr c2 1 c1;
SUBC> pred 20.
```
.
. (same regression output as above)
.

Predicted Values

```
    Fit   StDev Fit        95.0% CI              95.0% PI
  4.8692     0.0855    ( 4.6858,  5.0526)    ( 4.1195,  5.6189)
```

The value given in the column labeled Fit is the value of "gas consumption" given by the regression equation, when "degree-days" is equal to 20. The same output could be obtained by selecting **Stat ➤ Regression ➤ Regression**, and using the Options button to specify predicted values.

The TABLE Command

The **TABLE** command prints one-way, two-way, and multi-way tables. The format for the command is

```
TABLE the data classified by C...C
```

In the data set EG02-23.MTW, we have stored the information about 2900 hospital patients. For each patient, we have information about the patient's survival (in C1), the hospital (in C2), and the condition before surgery (in C3).

```
MTB > info
```

Information on the Worksheet

```
Column  Count  Name
T C1    2900   Outcome
T C2    2900   Hospital
T C3    2900   Condition
```

The **TABLE** command is conveniently used to determine the number of patients at hospital A and B that survived.

```
MTB > table 'outcome' 'hospital'
```

Tabulated Statistics

```
Rows: Outcome     Columns: Hospital

              A        B       All

Died         63       16        79
Survived   2037      784      2821
All        2100      800      2900

Cell Contents --
          Count
```

The **TABLE** command has a variety of subcommands available; more than one subcommand can be used at once. To obtain a list of possible subcommands, the **HELP** command can be used. To obtain information about any of the subcommands, the **HELP** command can be used with the command and subcommand name. Below we use the **TABLE** command with the **COLPERCENTS** subcommand. If no subcommand is used, counts are provided by the **TABLE** command. After the table is printed, the cell contents are identified. In the table below, the cell contents are expressed as percents of the column values.

```
MTB > Table 'outcome' 'hospital';
SUBC> colpercent.
```

Tabulated Statistics

```
Rows: Outcome     Columns: Hospital

              A        B       All

Died       3.00     2.00      2.72
Survived  97.00    98.00     97.28
All      100.00   100.00    100.00

Cell Contents --
          % of Col
```

The two-way table above indicates that Hospital A loses 3% of its surgery patients, and Hospital B loses only 2%. The same output can be obtained by selecting **Stat ➤ Tables ➤ Cross tabulation** from the menu, selecting the variables to be classified and the type of display.

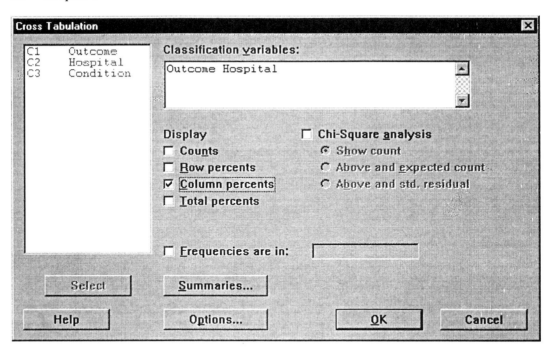

If three columns are specified in the **TABLE** command, three-way tables will be produced. These are essentially two-way tables for each value of the third variable specified. Below we examine three-way tables and see that Hospital A beats Hospital B for patients in good condition (only 1% died compared with 1.33% in Hospital B). Hospital A wins again for patients in poor condition losing 3.8% to Hospital B's 4%. This is illustrated using the **TABLE** command with the **COLPERCENTS** subcommand.

```
MTB > Table 'Outcome' 'Hospital' 'Condition';
SUBC>    ColPercents.
```

Tabulated Statistics

```
Control: Conditio = Good
Rows: B      All
Outcome      Columns: Hospital

                 A
Died         1.00      1.33      1.17
Survived    99.00     98.67     98.83
All        100.00    100.00    100.00
```

```
Control: Conditio = Poor
  Rows: Outcome      Columns: Hospital

                 A          B         All

   Died        3.80       4.00       3.82
   Survived   96.20      96.00      96.18
   All       100.00     100.00     100.00

   Cell Contents --
                      % of Col
```

EXERCISES

2.4 Manatees are large, gentle sea creatures that live along the Florida coast. Many manatees are killed or injured by powerboats. Here and in EX02-04.MTW are data on powerboat registrations (in thousands) and the number of manatees killed by boats in Florida in the years 1977–1990.

Year	Powerboat registrations (1000)	Manatees killed	Year	Powerboat registrations (1000)	Manatees killed
1977	447	13	1984	559	34
1978	460	21	1985	585	33
1979	481	24	1986	614	33
1980	498	16	1987	645	39
1981	513	24	1988	675	43
1982	512	20	1989	711	50
1983	526	15	1990	719	47

(a) We want to examine the relationship between number of powerboats and number of manatees killed by boats. Which is the explanatory variable?

(b) Make a **PLOT** of these data. What does the scatterplot show about the relationship between these variables? Describe the direction of the relationship. Are the variables positively or negatively associated? Describe the form of the relationship. Is it linear?

(c) Describe the strength of the relationship. Can the number of manatees killed be predicted accurately from powerboat registrations? If powerboat registrations remained constant at 716,000, **PREDICT** the number of manatees that would be killed by boats each year.

2.6 How does the fuel consumption of a car change as its speed increases? Here and in EX02-06.MTW are data for a British Ford Escort. Speed is measured in kilometers per hour, and fuel consumption is measured in liters of gasoline used per 100 kilometers traveled.

Speed (km/h)	Fuel used (liters/100 km)	Speed (km/h)	Fuel used (liters/100 km)
10	21.00	90	7.57
20	13.00	100	8.27
30	10.00	110	9.03
40	8.00	120	9.87
50	7.00	130	10.79
60	5.90	140	11.77
70	6.30	150	12.83
80	6.95		

(a) Make a **PLOT** of the data. (Which is the explanatory variable?)

(b) Describe the form of the relationship. Why is it not linear? Explain why the form of the relationship makes sense.

(c) It does not make sense to describe the variables as either positively associated or negatively associated. Why?

(d) Is the relationship reasonably strong or quite weak? Explain your answer.

2.7 Metabolic rate, the rate at which the body consumes energy, is important in studies of weight gain, dieting, and exercise. The table below and TA02-02.MTW give data on the lean body mass and resting metabolic rate for 12 women and 7 men who are subjects in a study of dieting. Lean body mass, given in kilograms, is a person's weight leaving out all fat. Metabolic rate is measured in calories burned per 24 hours, the same calories used to describe the energy content of foods. The researchers believe that lean body mass is an important influence on metabolic rate.

(a) Make a **PLOT** of the data for the female subjects. Which is the explanatory variable?

(b) Is the association between these variables positive or negative? What is the form of the relationship? How strong is the relationship?

Subject	Sex	Mass	Rate	Subject	Sex	Mass	Rate
1	M	62.0	1792	11	F	40.3	1189
2	M	62.9	1666	12	F	33.1	913
3	F	36.1	995	13	M	51.9	1460
4	F	54.6	1425	14	F	42.4	1124
5	F	48.5	1396	15	F	34.5	1052
6	F	42.0	1418	16	F	51.1	1347
7	M	47.4	1362	17	F	41.2	1204
8	F	50.6	1502	18	M	51.9	1867
9	F	42.0	1256	19	M	46.9	1439
10	M	48.7	1614				

(c) Now add the data for the male subjects to your graph, using the **SYMBOL** subcommand. Does the type of relationship that you observed in (b) hold for men also? How do the male subjects as a group differ from the female subjects as a group?

2.9 Are hot dogs that are high in calories also high in salt? The calories and salt content (measured as milligrams of sodium) for each of 17 brands of meat hot dogs are given below and in EX02-09.MTW.

Calories	173	191	182	190	172	147	146	139	175	136
	179	153	107	195	135	140	138			
Sodium	458	506	473	545	496	360	387	386	507	393
	405	372	144	511	405	428	339			

(a) Use the **PLOT** command to make a scatterplot of the amount of sodium in a hot dog of each brand versus the number of calories. Describe the main features of the relationship.

(b) Compute two least-squares **REGRESSION** lines: one calculated using all of the observations, the other omitting the brand of veal hot dogs that is an outlier in both variables measured. Draw both regression lines on the scatterplot.

2.11 There is some evidence that drinking moderate amounts of wine helps prevent heart attacks. Table 2.3 in BPS and TA02-03.MTW give data on yearly wine consumption (liters of alcohol from drinking wine, per person) and yearly deaths from heart disease (deaths per 100,000 people) in 19 developed nations.

(a) Make a **PLOT** that shows how national wine consumption helps explain heart disease death rates. Describe the form of the relationship. Is there a linear pattern? How strong is the relationship?

(b) Find the **CORRELATION** for these variables. About what percent of the variation among countries in heart disease death rates is explained by the straight-line relationship with wine consumption?

(c) Find the least-squares **REGRESSION** line for predicting heart disease death rate from wine consumption, calculated from the data in Table 2.3. Use this equation to predict the heart disease death rate in another country where adults average 4 liters of alcohol from wine each year.

2.14 Table 1.6 in BPS and TA01-06.MTW give data for the states. We might expect that states with less educated populations would pay their teachers less, perhaps because these states are poorer.

(a) Make a **PLOT** of average teachers' pay against the percent of state residents who are not high school graduates. Take the percent with no high school degree as the explanatory variable.

(b) The plot shows a weak negative association between the two variables. Why do we say that the association is negative? Why do we say that it is weak?

(c) Circle on the plot the point for the state your school is in.

(d) There is an outlier at the upper left of the plot. Which state is this?

(e) We wonder about regional patterns. There is a relatively clear cluster of nine states at the lower right of the plot. These states have many residents who are not high school graduates and pay low salaries to teachers. Which states are these? Are they mainly from one part of the country?

2.15 Data analysts often look for a simple *transformation* of data that simplifies the overall pattern. Here is an example of how transforming the response variable can simplify the pattern of a scatterplot. The population of Europe grew as follows between 1750 and 1950. Enter into a worksheet the following data.

Year	1750	1800	1850	1900	1950
Populations (millions)	125	187	274	423	594

(a) Make a **PLOT** of population against year. Briefly describe the pattern of Europe's growth.

(b) Now use the **LET** command or Minitab's calculator to take the logarithm of the population in each year. **PLOT** the logarithms against year using the following Minitab commands. What is the overall pattern on this plot?

2.23 A food industry group asked 3368 people to guess the number of calories in each of several common foods. Here is a table of the average of their guesses and the correct number of calories. The data are also stored in TA02-04.MTW.

Food	Guessed calories	Correct calories
8 oz. whole milk	196	159
5 oz. spaghetti with tomato sauce	394	163
5 oz. macaroni with cheese	350	269
One slice wheat bread	117	61
One slice white bread	136	76
2-oz. candy bar	364	260
Saltine cracker	74	12
Medium-size apple	107	80
Medium-size potato	160	88
Cream-filled snack cake	419	160

(a) We think that how many calories a food actually has helps explain people's guesses of how many calories it has. With this in mind, make a **PLOT** of these data.

(b) Find the **CORRELATION**. Explain why your *r* is reasonable based on the scatterplot.

(c) The guesses are all higher than the true calorie counts. Does this fact influence the correlation in any way? How would *r* change if every guess were 100 calories higher?

(d) The guesses are much too high for spaghetti and snack cake. Circle these points on your scatterplot. Calculate *r* for the other eight foods, after deleting these two points from the worksheet. Explain why *r* changed in the direction that it did.

2.33 Here and in EX02-33.MTW are Professor Moore's times (in minutes) to swim 2000 yards and his pusle rate after swimming (in beats per minute) for 23 sessions in the pool:

Time	34.12	35.72	34.72	34.05	34.13	35.72	36.17	35.57
Pulse	152	124	140	152	146	128	136	144
Time	35.37	35.57	35.43	36.05	34.85	34.70	34.75	33.93
Pulse	148	144	136	124	148	144	140	156
Time	34.60	34.00	34.35	35.62	35.68	35.28	35.97	
Pulse	136	148	148	132	124	132	139	

(a) Make a **PLOT** with time as the explanatory variable. Describe the form and strength of the relationship.

(b) Find the least-squares **REGRESSION** line.

(c) The next day's time is 34.30 minutes. **PREDICT** the professor's pulse rate. In fact, his pulse rate was 152. How accurate is your prediction?

(d) Suppose you were told only that the pulse rate was 152. You now want to predict swimming time. Find the equation of the least-squares regression line that is appropriate for this purpose. What is your prediction, and how accurate is it?

(e) Explain clearly, to someone who knows no statistics, why there are two different regression lines.

2.37 Exercise 2.23 and TA02-04.MTW give data on the true calories in 10 foods and the average guesses made by a large group of people

(a) Make a **PLOT** suitable for predicting guessed calories from true calories. Circle the points for spaghetti and snack cake on your plot. These points lie outside the linear pattern of the other eight points.

(b) Find the least-squares **REGRESSION** line of guessed calories on true calories. Do this twice, first for all 10 data points and then leaving out spaghetti and snack cake.

(c) Plot by hand both lines on your graph. (Make one dashed so you can tell them apart.) Are spaghetti and snack cake, taken together, influential observations? Explain your answer.

2.45 Investors ask about the relationship between returns on investments in the United States and on investments overseas. Here and in TA02-07.MTW are data on the total returns on U.S. and overseas common stocks over a 27-year period. (The total return is change in price plus any dividends paid converted into U.S. dollars. Both returns are averages over many individual stocks.)

Year	Overseas % return	U.S. % return	Year	Overseas % return	U.S. % return
1971	29.6	14.6	1985	56.2	31.6
1972	36.3	18.9	1986	69.4	18.6
1973	−14.9	−14.8	1987	24.6	5.1
1974	−23.2	−26.4	1988	28.3	16.6
1975	35.4	37.2	1989	10.5	31.5
1976	2.5	23.6	1990	−23.4	−3.1
1977	18.1	−7.4	1991	12.5	30.4
1978	32.6	6.4	1992	−11.8	7.6
1979	4.8	18.2	1993	32.9	10.1
1980	22.6	32.3	1994	6.2	1.3
1981	−2.3	−5.0	1995	11.2	37.6
1982	−1.9	21.5	1996	6.1	23.0
1983	23.7	22.4	1997	2.1	33.4
1984	7.4	6.1			

(a) Make a **PLOT** suitable for predicting overseas returns from U.S. returns.

(b) Find the **CORRELATION** and r^2. Describe the relationship between U.S. and overseas returns in words, using r and r^2 to make your description more precise.

(c) Find the least-squares **REGRESSION** line of overseas returns on U.S. returns. Draw the line on the scatterplot.

(d) In 1997, the return on U.S. stocks was 33.4%. Use the regression line to **PREDICT** the return on overseas stocks. The actual overseas return was 2.1%. Are you confident that predictions using the regression line will be quite accurate? Why?

(e) Circle the point that has the largest residual (either positive or negative). What year is this? Are there any points that seem likely to be very influential?

2.46 TA02-08.MTW contains four sets of bivariate data prepared by the statistician Frank Anscombe to illustrate the dangers of calculating without first plotting the data. The data are also given below.

					Data Set A						
x	10	8	13	9	11	14	6	4	12	7	5
y	8.04	6.95	7.58	8.81	8.33	9.96	7.24	4.26	10.84	4.82	5.68

					Data Set B						
x	10	8	13	9	11	14	6	4	12	7	5
y	9.14	8.14	8.74	8.77	9.26	8.10	6.13	3.10	9.13	7.26	4.74

					Data Set C						
x	10	8	13	9	11	14	6	4	12	7	5
y	7.46	6.77	12.74	7.11	7.81	8.84	6.08	5.39	8.15	6.42	5.73

					Data Set D						
x	8	8	8	8	8	8	8	8	8	8	19
y	6.58	5.76	7.71	8.84	8.47	7.04	5.25	5.56	7.91	6.89	12.50

(a) Calculate the **CORRELATION** and the least-squares **REGRESSION** line for all three data sets and verify that they agree.

(b) **PLOT** each of the three data sets and draw the regression line on each of the plots.

(c) In which of the four cases would you be willing to use the fitted regression line to describe the dependence of y on x? Explain your answer in each case.

2.49 Exercise 2.45 examined the relationship between returns on U.S. and overseas stocks. The data are given in TA02-07.MTW. Investors also want to know what typical returns are and how much year-to-year variability (called *volatility* in finance) there is. Regression and correlation do not answer these questions.

(a) Use the **DESCRIBE** command to find the five-number summaries for both U.S. and overseas returns. Make side-by-side **BOXPLOTS** to compare the two distributions.

(b) Were returns generally higher in the United States or overseas during this period? Explain your answer.

(c) Were returns more volatile (more variable) in the United States or overseas during this period? Explain your answer.

2.53 The number of people living on American farms has declined steadily during the 20th century. Here and in EX02-53.MTW are data on the farm population (millions of persons) from 1935 to 1980.

Year	1935	1940	1945	1950	1955	1960	1965	1970	1975	1980
Population	32.1	30.5	24.4	23.0	19.1	15.6	12.4	9.7	8.9	7.2

(a) Make a **PLOT** of these data and find the least-squares **REGRESSION** line of farm population on year.

(b) According to the regression line, how much did the farm population decline each year on the average during this period? What percent of the observed variation in farm population is accounted for by linear change over time?

(c) **PREDICT** the number of people living on farms in 1990. Is this result reasonable? Why?

2.70 Here and in EX2-70.MTW are data from eight high schools on smoking among students and among their parents:

	Neither parent smokes	One parent smokes	Both parents smoke
Student does not smoke	1168	1823	1380
Student smokes	188	416	400

(a) Use the **TABLE** command to determine how many students these data describe.

(b) What percent of these students smoke? Should you use the **ROWPERCENTS** or **COLPERCENTS** subcommand to answer this question?

(c) Give the marginal distribution of parents' smoking behavior, both in counts and in percents. Use both the **ROWPERCENTS** and the **COLPERCENTS** subcommands.

2.77 Whether a convicted murderer gets the death penalty seems to be influenced by the race of the victim. The following three-way table and EX02-77.MTW classify 326 cases in which the defendant was convicted of murder. The three variables are the defendant's race, the victim's race, and whether the defendant was sentenced to death.

	White Defendant			Black Defendant	
	White victim	Black victim		White victim	Black victim
Death	19	0	Death	11	6
Not	132	9	Not	52	97

(a) Form a two-way **TABLE** of defendants' races by death penalty.

(b) Show that Simpson's paradox holds: A higher percent of white defendants are sentenced to death overall, but for both black and white victims a higher percent of black defendants are sentenced to death.

(c) Use the data to explain why the paradox holds in language that a judge could understand.

2.87 Upper Wabash Tech has two professional schools, business and law. Here is a three-way table of applicants to these professional schools, categorized by sex, school, and admission decision. The data are also provided in EX02-87.MTW.

	Business			Law	
	Admit	Deny		Admit	Deny
Male	480	120	Male	10	90
Female	180	20	Female	100	200

(a) Make a two-way **TABLE** of sex by admission decision for the combined professional schools by summing entries in the three-way table.

(b) Use the **ROWPERCENTS** subcommand to compute separately the percents of male and female applicants admitted from your two-way table. Upper Wabash Tech's professional schools admit a higher percent of male applicants than of female applicants.

(c) Now compute separately the percents of male and female applicants admitted by the business school and by the law school. Each school admits a higher percent of female applicants.

(d) Explain carefully, as if speaking to a skeptical reporter, how it can happen that Upper Wabash appears to favor males when each school individually favors females.

2.96 Ecologists collect data to study nature's patterns. Table 2.11 in BPS and TA02-11.MTW give data on the mean number of seeds produced in a year by several common tree species and the mean weight (in milligrams) of the seeds produced. (Some species appear twice because their seeds were counted in two locations.) We might expect that trees with heavy seeds produce fewer of them, but what is the form of the relationship?

(a) Make a **PLOT** showing how the weight of tree seeds helps explain how many seeds the tree produces. Describe the form, direction, and strength of the relationship.

(b) When dealing with sizes and counts, the logarithms of the original data are often the "natural" variables. Use the **LET** command or Minitab's calculator to obtain the logarithms of both the seed weights and the seed counts in Table 2.11. Make a new scatterplot using these new variables. What are the form, direction, and strength of the relationship?

2.99 Table 1.6 and TA01-06.MTW give data about education in the states. Examine the relationship between the median SAT verbal and mathematics scores as follows.

(a) You want to predict a state's SAT math score from its verbal score. Find the least-squares **REGRESSION** line for this purpose. You learn that a state's median verbal score the following year was 455. **PREDICT** its median math score.

(b) **PLOT** the residuals from your regression against the SAT verbal score. One state is an outlier. What state is this? Does this state have a median math score higher or lower than would be predicted on the basis of its median verbal score?

2.102 In the mid-1970s, a medical study contacted randomly chosen people in a district in England. Here and in EX02-102.MTW are data on the 1314 women contacted who were either current smokers or who had never smoked. The table classifies these women by their smoking status and age at the time of the survey and whether they were still alive 20 years later.

	Age 18 to 44		Age 45 to 64		Age 65+	
	Smoker	Non-smoker	Smoker	Non-smoker	Smoker	Non-smoker
Dead	19	13	78	52	42	165
Alive	269	327	167	147	7	28

(a) From these data, make a two-way **TABLE** of smoking (yes or no) by dead or alive. Use the **COLPERCENTS** subcommand to find the percent of the smokers who stayed alive for 20 years. What percent of the non-smokers survived? It seems surprising that a higher percent of smokers stayed alive.

(b) The age of the women at the time of the study is a lurking variable. Show with a three-way **TABLE** (and **COLPERCENTS**) that within each of the three age groups in the data, a higher percent of non-smokers remained alive 20 years later. This is another example of Simpson's paradox.

(c) The study authors give this explanation: "Few of the older women (over 65 at the original survey) were smokers, but many of them had died by the time of follow-up." Verify the explanation with a two-way **TABLE** (with **ROWPERCENTS**) of smoking and age.

Chapter 3
Producing Data

Commands to be covered in this chapter:

```
SAMPLE K rows from C...C put into C...C
SET data into C
SORT C [carry along C...C] put into C [and C...C]
UNSTACK (C...C) into (E...E) ... (E...E)
```

The SAMPLE Command

The **SAMPLE** command can be used to select a simple random sample from a population. The format for the command is

```
SAMPLE K rows from C...C, put into C...C
```

The constant K designates the sample size. K may be any integer that is less than or equal to the size of the input column. The command selects K rows (without replacement) at random from the specified columns. The rows are sampled in random order. If rows are sampled from several columns at once, the same rows are selected from each.

Suppose that a sample of 40 is to be selected from a population of 1000 individuals for the purpose of conducting a survey. The following Minitab commands could be used to select the sample. For illustration, we will label the individuals with identification numbers from 1 to 1000. A consecutive list of integers following the **SET** command may be abbreviated using a colon. The sequence 6, 7, 8, 9, 10 is abbreviated as 6:10. Below we enter the numbers 1 to 1000 into C1 and then select 40 individuals to be put into C2 using the **SAMPLE** command.

```
MTB > set c1
DATA> 1:1000
DATA> end
MTB > sample 40 c1 c2
MTB > name c2 'sample'
MTB > print c2
```

Data Display

```
sample
    722    753    439    929    738    723    446    212    413    139
    178    639    187    185    646    746    847    963    337    500
    594    908    733     32    806    936    490    554    491    955
    809    899    556    123    319    544    484    611    538    341
```

The identification numbers listed above refer to individuals in the population. Once a sample is selected, a population list is needed to determine which individuals are to be included in the sample and will be asked to answer survey questions.

The sample could also be selected by choosing **Calc ➤ Random Data ➤ Sample From Columns** from the menu and filling in the dialog box as shown.

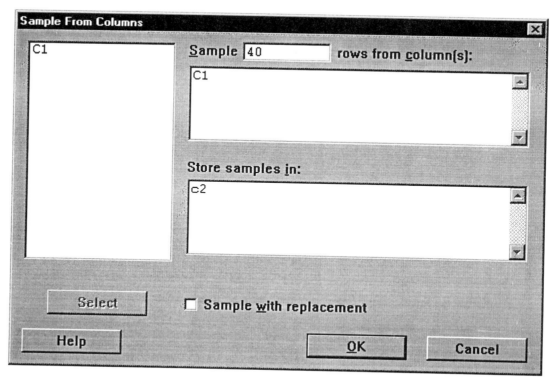

The SORT Command

The samples above were selected in random order. It may be convenient to sort these numbers. This is easily done using the **SORT** command. The **SORT** command orders the data in a column in numerical sequence. The format for the **SORT** command is as follows:

```
SORT C [carry along C...C] put into C [and C...C]
```

The identification numbers for the individuals to be included in the above random sample (C2) can be sorted and put into another column (C3) using the command:

```
MTB > sort c2 c3
MTB > name c3 'sort'
MTB > print c3
```

Data Display

```
sort
     32    123    139    178    185    187    212    319    337    341
    413    439    446    484    490    491    500    538    544    554
    556    594    611    639    646    722    723    733    738    746
    753    806    809    847    899    908    929    936    955    963
```

The second column specified in the **SORT** can be the same as the first column specified. In this case, the sorted data would simply replace the unsorted data.

The **SORT** command is available by selecting **Manip ➤ Sort** from the menu. In the dialog box, specify the columns you want to sort and where you'd like the result stored. If the descending box is left unchecked, sorting will be done from lowest to highest.

Both the **SAMPLE** and the **SORT** commands will work with text as well as numeric data. In this case, the **SORT** command will alphabetize. On the following page, we selected a sample of five clients to interview from the client list given in Example 3.4 of BPS. The client list is available in EG03-04.MTW.

Sort dialog

```
C1    Population
C2    sample
```

Sort column(s): sample

Store sorted column(s) in: c3

Sort by column: sample ☐ Descending
Sort by column: _____ ☐ Descending
Sort by column: _____ ☐ Descending
Sort by column: _____ ☐ Descending

[Select] [Help] [OK] [Cancel]

Data Display

```
Client
  A-1 Plumbing              Accent Printing
  Action Sport Shop         Anderson Construction
  Bailey Trucking           Balloons Inc.
  Bennett Hardware          Best's Camera Shop
  Blue Print Specialties    Central Tree Service
  Classic Flowers           Computer Answers
  Darlene's Dolls           Fleisch Realty
  Hernandez Electronics     JL Records
  Johnson Commodities       Keiser Construction
  Liu's Chinese Restaurant  MagicTan
  Peerless Machine          Photo Arts
  River City Books          Riverside Tavern
  Rustic Boutique           Satellite Services
  Scotch Wash               Sewer's Center
  Tire Specialties          Von's Video Store
```

```
MTB > sample 5 c1 c2
MTB > print c2
```

Data Display

```
Sample
   Johnson Commodities      Sewer's Center
   Bailey Trucking          Keiser Construction
   Tire Specialties

MTB > sort c2 c2
MTB > print c2
```

Data Display

```
Sample
   Bailey Trucking          Johnson Commodities
   Keiser Construction      Sewer's Center
   Tire Specialties
```

Randomization in Experiments

The **SAMPLE** command also can be used to randomly select treatment groups in an experiment. In example 3.13 in BPS, 60 family residences will participate in an experiment to reduce energy use. The electric utility company will randomly assign 20 residences to each of the three treatments. By selecting **Calc ➤ Make Patterned Data ➤ Simple Set of Numbers** from the menu, we can enter the numbers 1 through 60 into a Minitab worksheet. To enter the treatment names into column 2, select **Calc ➤ Make Patterned Data ➤ Text values** and fill out the dialog box as illustrated.

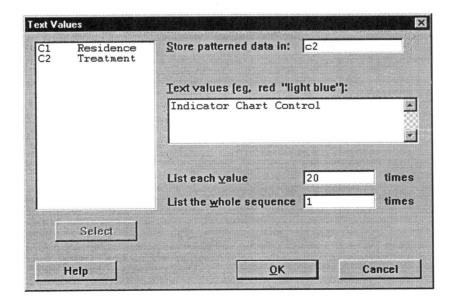

To make the assignment of treatments to residences random, we can use the **SAMPLE** command (on either C1 or C2) to randomize the order of the list. The window to the right shows the treatments listed in a random order. This was obtained with the Minitab command:

```
MTB > sample c2 c2
```

If the command

```
MTB > sample c1 c1
```

is used instead, the list of residences would be in a random order.

	C1 Residence	C2-T Treatment
1	1	Control
2	2	Chart
3	3	Chart
4	4	Indicator
5	5	Control
6	6	Indicator
7	7	Indicator
8	8	Chart
9	9	Indicator
10	10	Indicator
11	11	Chart
12	12	Control

The **SAMPLE** command also can be used to select treatment groups for more complicated experimental designs. Consider Example 3.10 of BPS. This experiment investigated the effects of repeated exposure to an advertising message. All subjects will view a 40-minute television program that includes ads for a 35 mm camera. Some subjects will see a 30-second commercial; others, a 90-second version. The same commercial will be repeated either 1, 3, or 5 times during the program. This experiment has two factors: length of commercial, with 2 levels, and repetitions, with 3 levels. The 6 combinations of one level of each factor form 6 treatments. Below we create a Minitab worksheet listing the possible treatments and subject identification numbers for 18 subjects. Parentheses are again used as an abbreviation for a repeat factor and a colon is used to abbreviate a list of integers.

```
MTB > set c1
DATA> 9(1) 9(2)
DATA> set c2
DATA> 3(1) 3(2) 3(3) 3(1) 3(2) 3(3)
MTB > set c3
DATA> 1:30
DATA> end
MTB > name c1 'length'
MTB > name c2 'reps'
MTB > name c3 'subjects'
MTB > print c1-c3
```

Data Display

ROW	length	reps	subjects
1	1	1	1
2	1	1	2
3	1	1	3
4	1	2	4
5	1	2	5
6	1	2	6
7	1	3	7
8	1	3	8
9	1	3	9
10	2	1	10
11	2	1	11
12	2	1	12
13	2	2	13
14	2	2	14
15	2	2	15
16	2	3	16
17	2	3	17
18	2	3	18

The worksheet above illustrates the experimental design that will be used, that is, 3 subjects in each of 6 experimental groups. The assignment of subjects into groups can be randomized using the **SAMPLE** command to randomly assign subjects to the 6 treatments. The command

```
MTB > sample 18 c3 c3
```

will randomly assign treatments to subjects.

76   Chapter 3

```
MTB > print c1-c3
```

Data Display

```
ROW    length    reps    subjects
 1        1        1        15
 2        1        1         1
 3        1        1        14
 4        1        2        17
 5        1        2        16
 6        1        2         2
 7        1        3         4
 8        1        3        10
 9        1        3         7
10        2        1        13
11        2        1         3
12        2        1         9
13        2        2        12
14        2        2        11
15        2        2         8
16        2        3        18
17        2        3         5
18        2        3         6
```

Note that the subjects are now arranged in a random order so that subjects 15, 1, and 14 are included in the first group (length level = 1, repetitions level = 1), subjects 17, 16, and 2 are included in the second group (length level = 1, repetitions level = 2), etc.

The UNSTACK Command

This command separates one or more columns into several blocks of columns. For most applications, the subcommand **SUBSCRIPTS** is needed. The rows with the smallest subscript are stored in the first block, the rows with the second smallest subscript in the second block, and so on. If you do not use the subcommand, each row is stored in a separate block. The values in the subscripts column must be integers. The format for the command is

```
UNSTACK (C...C) into (E...E) ... (E...E)
```

As described in Chapter 1, the command also can be used by selecting Data can also be stacked by selecting **Manip ➤ Stack/Unstack ➤ Unstack Columns** from the menu. This command is useful for separating experimental units for a randomized block experimental design. Consider, for example, the randomized block design outline in Figure 3.6 of BPS. The blocks consist of male and female subjects, while the treatments are three therapies for cancer. The following worksheet contains subject identification numbers for 30 subjects in the first column and codes to identify the sex of the subject (1 = male, 2 = female) in the second column. The **UNSTACK** command is used to separate the subjects into these two groups.

```
MTB > info
```

Information on the Worksheet

```
COLUMN      NAME        COUNT
C1          subject     30
C2          sex         30

CONSTANTS USED: NONE

MTB > print c1 c2

 ROW    subject    sex

  1        1        1
  2        2        2
  3        3        1
  4        4        2
  .        .        .
  .        .        .
  .        .        .
 29       29        1
 30       30        2

MTB > unstack 'subject' into c3 c4;
SUBC> subs 'sex'.
MTB > name c3 'male'
MTB > name c4 'female'
```

```
MTB > print c3 c4
```

Data Display

ROW	male	female
1	1	2
2	3	4
3	6	5
4	8	7
5	9	11
6	10	13
7	12	15
8	14	16
9	17	20
10	18	23
11	19	24
12	21	26
13	22	27
14	25	28
15	29	30

Note that the male subjects (sex = 1) were stored in the first block and the female subjects were stored in the second block. To complete the experimental design, we list the treatments in C5 and then randomize the data in each block (C3 and C4).

```
MTB > set c5
DATA> 5(1) 5(2) 5(3)
DATA> end
MTB > name c5 'treat'
MTB > sample 15 c3 c3
MTB > sample 15 c4 c4
```

```
MTB > print c3-c5
```

Data Display

```
ROW    male   female   treat

 1      8      16       1
 2      3      30       1
 3     29       4       1
 4      1      15       1
 5     19      26       1
 6      6      28       2
 7     22       7       2
 8     12       2       2
 9     21      27       2
10     14      24       2
11     18      20       3
12     25      13       3
13      9      23       3
14     10       5       3
15     17      11       3
```

The **SAMPLE** command has randomly assigned subjects 8, 3, 29, 1, and 19 (males) and subjects 16, 30, 4, 15, and 26 (females) to treatment one. The individuals assigned to treatments two and three can be read off the above worksheet listing.

EXERCISES

3.7 A firm wants to understand the attitudes of its minority managers toward its system for assessing management performance. Below and in EX03-07.MTW is a list of all the firm's managers who are members of minority groups. Use the **SAMPLE** command to choose 6 to be interviewed in detail about the performance appraisal system.

Agarwal	Dewald	Huang	Puri
Anderson	Fernandez	Kim	Richards
Baxter	Fleming	Liao	Rodriguez
Bonds	Gates	Mourning	Santiago
Bowman	Goel	Naber	Shen
Castillo	Gomez	Peters	Vega
Cross	Hernandez	Plieg	Wang

3.8 Your class in ancient Ugaritic religion is poorly taught and wants to complain to the dean. The class decides to choose 4 of its members at random

80 Chapter 3

to carry the complaint. The class list appears below and in EX03-08.MTW. Choose an SRS of 4 using the **SAMPLE** command.

Anderson	Gupta	Patnaik
Aspin	Gutierrez	Pirelli
Bennett	Harter	Rao
Bock	Henderson	Rider
Breiman	Hughes	Robertson
Castillo	Johnson	Rodriguez
Dixon	Kempthorne	Sosa
Edwards	Laskowsky	Tran
Gonzalez	Liang	Trevino
Green	Olds	Wang

3.10 A club contains 30 student members and 10 faculty members. The students are

Abel	Fisher	Huber	Moran	Reinmann
Carson	Golomb	Jimenez	Moskowitz	Santos
Chen	Griswold	Jones	Neyman	Shaw
David	Hein	Kiefer	O'Brien	Thompson
Deming	Hernandez	Klotz	Pearl	Utts
Elashoff	Holland	Liu	Potter	Vlasic

and the faculty members are

| Andrews | Fernandez | Kim | Moore | Rabinowitz |
| Besicovitch | Gupta | Lightman | Phillips | Yang |

The student and faculty names are given in EX03-10.MTW. The club can send four students and two faculty members to a convention and decides to choose those who will go by random selection. Use the **SAMPLE** command (twice) to choose a stratified random sample of 4 students and 2 faculty members.

3.24 A manufacturer of specialty chemicals chooses 3 from each lot of 25 containers of a reagent to be tested for purity and potency. The control numbers stamped on the bottles in the current lot are given below and in EX03-24.MTW. Use the **SAMPLE** command to choose an SRS of 3 of these bottles.

A1096	A1097	A1098	A1101	A1108
A1112	A1113	A1117	A2109	A2211
A2220	B0986	B1011	B1096	B1101
B1102	B1103	B1110	B1119	B1137
B1189	B1223	B1277	B1286	B1299

3.28 A corporation employs 2000 male and 500 female engineers. A stratified random sample of 200 male and 50 female engineers gives each engineer 1 chance in 10 to be chosen. This sample design gives every individual in the population the same chance to be chosen in the sample. Suppose that the employee names are listed so that employees 1 through 2000 are male and employees 2001 through 2500 are female.

(a) Enter the numbers 1 though 2500 into a Minitab worksheet and use the **SAMPLE** command to select an SRS of size 250.

(b) Determine how many of the individuals selected in the sample are male and how many are female? It will be easier if you use the **SORT** command first.

(c) Explain how the stratified random sample is different from an SRS.

3.34 A chemical engineer is designing the production process for a new product. The chemical reaction that produces the product may have higher or lower yield, depending on the temperature and the stirring rate in the vessel in which the reaction takes place. The engineer decides to investigate the effects of combinations of two temperatures (50° C and 60° C) and three stirring rates (60 rpm, 90 rpm, and 120 rpm) on the yield of the process. Two batches of the feedstock will be processed at each combination of temperature and stirring rate.

(a) How many factors and treatments are there in this experiment? How many experimental units (batches of feedstock) does the experiment require?

(b) Outline in graphic form the design of an appropriate experiment.

(c) The randomization in this experiment determines which batches of the feedstock will be processed according to each treatment. Use Minitab to carry out the randomization. Clearly state the results.

3.37 Will providing child care for employees make a company more attractive to women, even those who are unmarried? You are designing an experiment to answer this question. You prepare recruiting material for two fictitious companies, both in similar businesses in the same location. Company A's brochure does not mention child care. There are two versions of Company B's material, identical except that one describes the company's on-site child-care facility. Your subjects are 40 unmarried women who are college seniors seeking employment. Each subject will read recruiting material for both companies and choose the one she would prefer to work for. You will give each version of Company B's brochure to half the women. You expect that a higher percentage of those who read the description that includes child care will choose Company B.

(a) Outline an appropriate design for the experiment.

(b) The names of the subjects appear below and in EX03-37.MTW. Use Minitab to do the randomization required by your design. List the subjects who will read the version that mentions child care.

Abrams	Danielson	Gutierrez	Lippman	Rosen
Adamson	Durr	Howard	Martinez	Sugiwara
Afifi	Edwards	Hwang	McNeill	Thompson
Brown	Fluharty	Iselin	Morse	Travers
Cansico	Garcia	Janle	Ng	Turing
Chen	Gerson	Kaplan	Quinones	Ullmann
Cortez	Green	Kim	Rivera	Williams
Curzakis	Gupta	Lattimore	Roberts	Wong

3.45 Twenty overweight females have agreed to participate in a study of the effectiveness of four reducing regimens, A, B, C, and D. The researcher first calculates how overweight each subject is by comparing the subject's actual weight with her "ideal" weight. The subjects and their excess weights in pounds are given below and in EX03-45.MTW.

Birnbaum	35	Hernandez	25	Moses	25	Smith	29
Brown	34	Jackson	33	Nevesky	39	Stall	33
Brunk	30	Kendall	28	Obrach	30	Tran	35
Dixon	34	Loren	32	Rodriguez	30	Wilansky	42
Festinger	24	Mann	28	Santiago	27	Williams	22

The response variable is the weight lost after 8 weeks of treatment. Because the initial amount overweight will influence the response variable, a block design is appropriate.

(a) Use the **SORT** command to arrange the subjects in order of excess weight. Use the **UNSTACK** command to form five blocks by grouping the four least overweight, then the next four, and so on.

(b) Use Minitab to do the required random assignment of subjects to the four reducing regimens separately within each block. Be sure to explain exactly which subjects get each of the four treatments.

3.53 Some medical researchers suspect that added calcium in the diet reduces blood pressure. Forty men with high blood pressure are willing to serve as subjects.

(a) Outline an appropriate design for the experiment, taking the placebo effect into account.

(b) The names of the subjects appear below and in EX03-53.MTW. Use Minitab to do the randomization required by your design, and list the subjects to whom you will give the drug.

Alomar	Denman	Han	Liang	Rosen
Asihiro	Durr	Howard	Maldonado	Solomon
Bennett	Edwards	Hruska	Marsden	Tompkins
Bikalis	Farouk	Imrani	Moore	Townsend
Chen	Fratianna	James	O'Brian	Tullock
Clemente	George	Kaplan	Ogle	Underwood
Cranston	Green	Krushchev	Plochman	Willis
Curtis	Guillen	Lawless	Rodriguez	Zhang

Chapter 4
Probability and Sampling Distributions

Commands to be covered in this chapter:

```
RANDOM K observations into C...C
TALLY the data in C...C
RMEAN of E...E put into C
XBARCHART [C E]
SCHART [C E]
```

The RANDOM Command

The **RANDOM** command is used to generate random numbers. The format for the command is

```
RANDOM K observations into C...C
```

where K is the number of random numbers to be generated and C...C are the column(s) in which the random numbers are to be stored. If no subcommands are used, **RANDOM** generates random numbers from the normal distribution, with a mean of 0 and a standard deviation of 1. The command can also be used to generate a random sequence of 0s and 1s with the probability of selecting a 1 equal to p. This is done with the subcommand

```
BERNOULLI trials with p = K
```

To generate the outcome from 100 coin tosses, we use the **RANDOM** command with the **BERNOULLI** subcommand or select **Calc ➤ Random Data ➤ Bernoulli** from the menu. In the dialog box fill in the number of rows to be generated, where they are to be stored, and the probability of success.

```
MTB > Random 100 C1;
SUBC>    Bernoulli .5.
```

Data Display

```
C1
    1  1  1  0  1  1  0  1  1  1  0  1
    1  1  0  1  1  0  1  0  0  0  0  0
    1  0  1  1  1  0  0  1  1  0  1  0
    1  1  1  1  1  1  1  1  1  0  1  1
    0  0  0  1  0  0  0  1  1  0  0  1
    0  0  0  1  0  0  0  0  0  1  0  1
    1  1  0  0  0  0  1  0  1  0  1  1
    1  0  0  0  1  1  0  1  1  0  0  0
    1  1  0  0
```

The **RANDOM** command also can be used with several columns at a time to do repeated sampling. To generate 20 replications of the above sample, specify 20 columns in the command.

```
MTB > random 100 c1-c20;
SUBC> bern .5.
```

The TALLY Command

The **TALLY** command produces and prints tables for each column named in the following format:

```
TALLY the data in C...C
```

If no subcommand is used, the output will contain frequency counts for every distinct value in the input column. The columns must contain integers or missing values (*). The **PERCENTS** subcommand gives percentages for each distinct value in the input column(s), starting at the smallest distinct value. The **COUNTS** subcommand gives frequency counts for each distinct value in each input column. This is the default action of **TALLY** if no subcommands are given. The **CUMCOUNTS** subcommand gives cumulative frequency counts for each distinct value in the input column(s), starting at the smallest distinct value. The **CUMPERCENTS** subcommand gives cumulative percentage values for each distinct value in the input column(s), starting at the smallest distinct value. The **ALL** subcommand produces all four of the **TALLY** functions. The **TALLY** command also can be used by selecting **Stat ➤ Tables ➤ Tally** from the menu.

The **TALLY** command is useful for summarizing the results of several replications. Below, we generate 20 samples of size 100 and summarize the results using the **PERCENTS** subcommand.

```
MTB > random 100 c1-c20;
SUBC> bern .5.
MTB > tally c1-c20;
SUBC> percents.
```

Summary Statistics for Discrete Variables

C1	Percent	C2	Percent	C3	Percent	C4	Percent
0	46.00	0	51.00	0	54.00	0	54.00
1	54.00	1	49.00	1	46.00	1	46.00

C5	Percent	C6	Percent	C7	Percent	C8	Percent
0	50.00	0	49.00	0	53.00	0	49.00
1	50.00	1	51.00	1	47.00	1	51.00

C9	Percent	C10	Percent	C11	Percent	C12	Percent
0	57.00	0	52.00	0	53.00	0	51.00
1	43.00	1	48.00	1	47.00	1	49.00

C13	Percent	C14	Percent	C15	Percent	C16	Percent
0	46.00	0	46.00	0	50.00	0	45.00
1	54.00	1	54.00	1	50.00	1	55.00

C17	Percent	C18	Percent	C19	Percent	C20	Percent
0	49.00	0	53.00	0	47.00	0	53.00
1	51.00	1	47.00	1	53.00	1	47.00

To observe the variability of the SRSs, we can display the results of the 20 replications in a histogram and describe the data. To do this, we first need to reenter the data.

```
MTB > set c21
DATA> 46   51   54   54   50   49   53   49   57   52
DATA> 53   51   46   46   50   45   49   53   47   53
DATA> end
MTB > name c21 'total'
MTB > hist c21
```

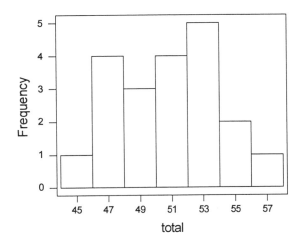

The DISCRETE Subcommand

We have used the **RANDOM** command to generate Bernoulli observations. The **DISCRETE** subcommand allows us to specify a discrete distribution. The values and corresponding probabilities are put into two columns and the following subcommand format is used to generate a random sample of observations.

```
DISCRETE dist. with values in C and probs in C
```

Any discrete distribution can be specified by putting the values and corresponding probabilities into two columns. We will simulate observations from the distribution of household sizes given below. First we enter the sizes and probabilities into a Minitab spreadsheet. These can either be entered into the Data window or using the session commands below.

Household size	1	2	3	4	5	6	7
Probability	.251	.321	.171	.154	.067	.022	.014

```
MTB > set c1
DATA> 1 2 3 4 5 6 7
MTB > set c2
DATA> .251 .321 .171 .154 .067 .022 .014
MTB > name c1 'size' c2 'prob'
```

Now, we can either enter the following session commands, or we can select **Calc ➤ Random Data ➤ Discrete** from the menu and fill in the dialog box as show.

```
MTB > random 50 c3;
SUBC> discrete c1 c2.

MTB > name c3 'random'
MTB > print c3
```

Data Display

```
random
    2   4   2   3   1   1   2   1   2   2   1   2
    3   1   5   1   4   3   2   4   2   3   4   2
    4   2   2   3   4   5   2   5   2   1   5   2
    1   2   3   2   3   1   5   2   3   4   1   4
    2   4
```

The UNIFORM Subcommand

To generate random numbers that are spread out uniformly between two numbers, the following subcommand format is used with the **RANDOM** command.

```
UNIFORM [continuous on the interval K to K]
```

The uniform distribution covers the interval from a to b. If you omit the arguments, then $a = 0$ and $b = 1$ are used. Below, we let $a = 5$ and $b = 10$ and generate 200 observations from the uniform distribution.

```
MTB > random 200 c1;
SUBC> unif 5 10.
MTB > hist c1
```

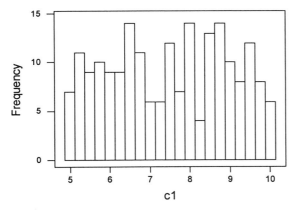

The NORMAL Subcommand

Observations from a normal distribution can be simulated with the **RANDOM** command and the **NORMAL** subcommand. The subcommand format is

```
NORMAL [mu = K [sigma = K]]
```

If you omit the arguments, then $\mu = 0$ and $\sigma = 1$ are used. In example 4.12 of BPS, we found that the DMS odor threshold for adults follows roughly a normal distribution with mean $\mu = 25$ micrograms per liter and standard deviation $\sigma = 7$ micrograms per liter. With this information, we can use the **RANDOM** command and the **NORMAL** subcommand to simulate many samples drawn from the population. To take 1000 samples each of size 10, we use the commands as follows.

```
MTB > random 10 c1-c1000;
SUBC> normal 25 7.
```

Selecting **Calc ➤ Random Data ➤ Normal** from the menu and filling in the dialog box will also give us random values simulated from a normal distribution.

The RMEAN Command

If we wish to find the 1000 sample mean thresholds $\bar{x}$, and make a histogram of these 1000 values, then it will be easier if we simulate 1000 rows and 10 columns Instead of 10 rows and 1000 columns.

```
MTB > RANDOM 1000 c1-c10;
SUBC> NORM 25 7.
```

The **RMEAN** command is the rowwise version of the command **MEAN**. It calculates and stores the arithmetic mean of the values in each row. Missing values are omitted from the calculation. The command format is as follows.

```
RMEAN of E...E put into C
```

The command also can be used by selecting **Calc ➤ Row Statistics** from the menu.

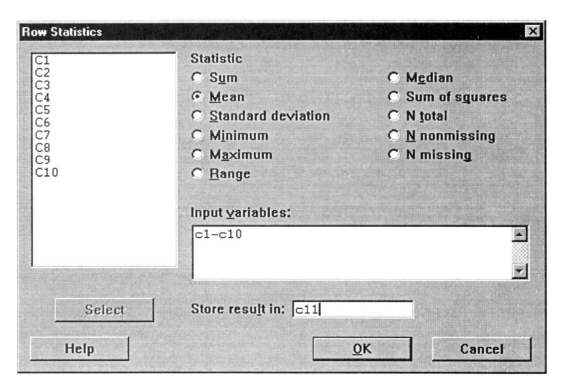

```
MTB > RMean c1-c10 c11
MTB > name c1 'X-bar'
```

A histogram of our 1000 values looks approximately normal with the center close to the population mean $\mu = 25$ and with a standard deviation that is much smaller than the population standard deviation $\sigma = 7$.

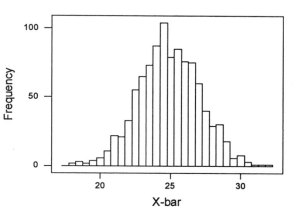

The Central Limit Theorem

Example 4.13 in BPS illustrates the Central Limit Theorem. We can do this using the **RANDOM** and **RMEAN** commands. This time we will generate observations from a population with an exponential distribution. This distribution is strongly right skewed. The commands below will generate 250 rows in 25 columns. In the column names X-bar(2) we calculate 250 sample means from 2 values, in the column names X-bar(5) we calculate 250 sample means from 5 values, and in the column named X-bar(25) we calculate 250 sample means from 25 values.

```
MTB > name c16 'X-bar(2)' c17 'X-bar(5)' c18 'X-bar(15)'
MTB > rmean c1 c2 'X-bar(2)'
MTB > rmean c1-c5 'X-bar(5)'
MTB > rmean c1-c15 'X-bar(15)'
MTB > hist c16-c18
```

The histograms below illustrate the right skewness of the original data and then samples means from 2, 5, and 25 observations. As *n* increases, the shape of the distributions, becomes more normal. The mean stays at $\mu = 1$, and the standard deviation decreases.

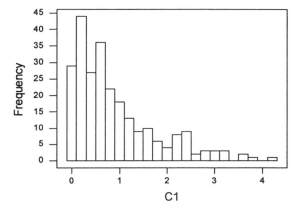
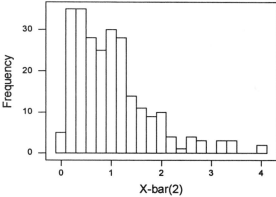

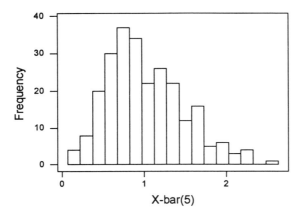
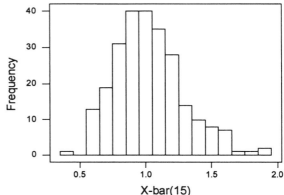

The XBACHART Command*

The **XBARCHART** command displays a separate $\bar{x}$ chart (a chart of sample means) for each variable specified by the following format.

    XBARCHART [C E]

E specifies the subgroups (also called samples). If E is a constant, say five, then Minitab takes the first 5 rows as the first sample, the second 5 rows as the second sample, and so on. If E is a column, then the subscripts in that column determine the subgroups. A new subgroup is formed every time the value in the subgroup column changes (this is different from the way most Minitab commands handle subscripts).

The mean of each subgroup is calculated. These means are plotted on the chart. In addition, a center line, an upper control limit (UCL) at 3σ above the center line, and a lower control limit (LCL) at 3σ below the center line are drawn on the chart.

A manufacturer of computer monitors measures the tension of fine wires behind the viewing screen. Four measurements are made every hour. The following data contain the measurements for 20 hours. The first four observations are from the first hour, the next four are from the second hour, etc. There are a total of 80 observations.

---

* The remainder of this chapter refers to material available on *The Basic Practice of Statistics* CD-ROM.

```
MTB > print c1

sample
```

239.9	299.9	352.8	249.3	257.7	305.3	183.2	238.5	233.5
270.8	240.6	294.5	255.3	257.3	261.7	241.1	180.6	302.6
317.4	253.7	267.4	287.3	225.1	179.7	260.8	287.9	249.9
292.5	226.0	290.0	334.1	323.1	305.4	269.2	335.0	323.9
292.3	259.1	371.6	255.8	227.1	297.6	315.8	247.5	290.7
202.9	298.3	258.6	299.0	318.1	259.0	287.0	264.8	232.9
293.2	186.0	272.9	193.1	303.7	381.6	329.1	287.1	261.9
220.7	311.1	312.9	310.5	308.0	288.6	161.1	241.0	222.8
306.8	277.2	294.3	265.8	297.3	313.8	361.9	284.3	

In the following $\bar{x}$ control chart, the mean and standard deviation are estimated from the data. This can be done either by selecting **Stat ➤ Control Charts ➤ Xbar** from the menu or with the following command.

```
MTB > xbar c1 4
```

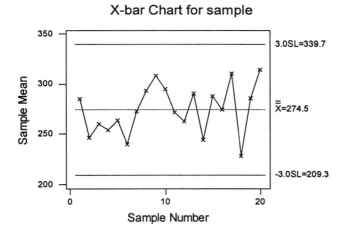

The process mean μ standard deviation σ were estimated above from the data. Alternatively, the parameters (μ and σ) may be specified from historical data with the following subcommands.

```
MU    = K
SIGMA = K
```

The value of μ is then used for the center line and σ is used in the calculations for the LCL and UCL. The default is that the upper and lower control limits (UCL and LCL) are 3σ above and below μ. The subcommand **SLIMITS** can be used to override the 3σ default and specify how many σ limits to use for LCL and UCL. The subcommand format is

94 *Chapter 4*

SLIMITS are K...K

The subcommand, SLIMITS 1 2 plots five lines: a center line, a line at 2σ above the center line, a line at σ above, a line at σ below the center line, and a line at 2σ below. The default lines, at 3σ limits, are not plotted.

Below we specify the mean and standard deviation for the process, and Minitab computes the mean and standard deviation for $\bar{x}$. These values are $\mu = 275$ and $\sigma = 43$ based on what is known about the process. These historical values can also be specified with subcommands or by filling in the values in the dialog box.

```
MTB > xbar c1 4;
SUBC> mu=275;
SUBC> sigma=43.
```

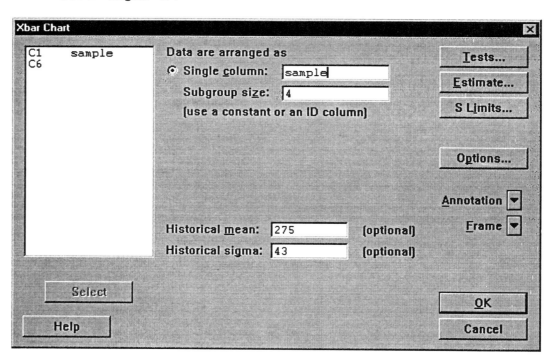

In practice, we must control the center of a process and it's variability. This is mostly commonly done with an *s* chart, a chart of standard deviations against time. The *s* chart can be produced by either selecting **Stat ➤ Control Charts ➤ S** from the menu or using the `SCHART` command. The command format is

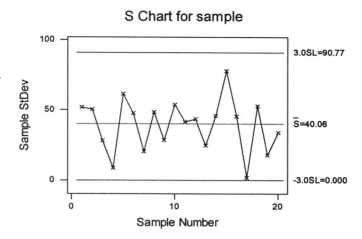

`SCHART [C E]`

The command is used the same way the `XBARCHART` is used, specifying a column with the samples and a sample size. The subcommand, `SCHART c1 4`, produces the displayed chart. Often, *s* minus 3 standard deviations gives a negative number. In this case, the lower control limit is plotted at 0.

Usually, the $\bar{x}$ chart and the *s* chart will be looked at together. We can produce both charts at once by selecting **Stat ➤ Control Charts ➤ Xbar S** from the menu. As usual, we have a choice of specifying or not specifying the historical values of μ and σ.

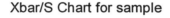

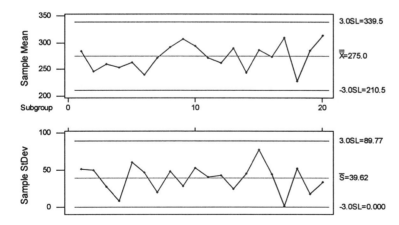

Out of Control Signals

The `XBACHART` and `SCHART` commands perform tests to identify out of control signals. Each test detects a specific pattern in the data plotted on the chart. The occurrence of a pattern suggests a special cause for the variation, one that should be investigated. The `TEST` subcommand format shown below can be used to select one or more of eight tests for special causes.

    TEST K...K

List the numbers of the tests you want on the subcommand. The subcommand, `TEST 1 3`, specifies that tests 1 and 3 are to be performed. `TEST 1:8` says do all eight tests. The tests can also be selected by clicking the Tests button on the dialog box and choosing from the Tests subdialog box.

When a point fails a test, it is marked with the test number on the plot. If a point fails more than one test, the number of the first test in your list is the number displayed on the plot. In addition, a summary table is displayed with complete information.

Tests

Tests For Special Causes (default definitions)
○ Perform all eight tests
◉ Choose specific tests to perform
    ☑ One point more than 3 sigmas from center line
    ☐ Nine points in a row on same side of center line
    ☐ Six points in a row, all increasing or all decreasing
    ☐ Fourteen points in a row, alternating up and down
    ☐ Two out of three points more than 2 sigmas from center line (same side)
    ☐ Four out of five points more than 1 sigma from center line (same side)
    ☐ Fifteen points in a row within 1 sigma of center line (either side)
    ☐ Eight points in a row more than 1 sigma from center line (either side)

[Help]   [OK]   [Cancel]

EXERCISES

4.11 The basketball player Shaquille O'Neal makes about half of his free throws during an entire season. Use the **RANDOM** command with the **BERNOULLI** subcommand to simulate 100 free throws shot independently by a player who has probability 0.5 of making each shot.

 (a) What percent of the 100 shots did he hit?

 (b) Examine the sequence of hits and misses. How long was the longest run of shots made? Of shots missed? (Sequences of random outcomes often show runs longer than our intuition thinks likely.)

4.13 A recent opinion poll showed that about 73% of married women agree that their husbands do at least their fair share of household chores. Suppose that this is true. Choosing a married woman at random then has probability 0.73 of getting one who agrees that her husband does his share. Use the **RANDOM** command and the **BERNOULLI** subcommand to simulate choosing many women independently.

 (a) Simulate drawing 20 women, then 80 women, then 320 women. What proportion agree in each case? We expect (but because of chance variation we can't be sure) that the proportion will be closer to 0.73 in longer runs of trials.

 (b) Simulate drawing 20 women 10 times and record the percents in each trial who agree. Then simulate drawing 320 women 10 times and again record the percents. Which set of 10 results is less variable? We expect the results of 320 trials to be more predictable (less variable) than the results of 20 trials. That is "long-run regularity" showing itself.

4.40 Let us illustrate the idea of a sampling distribution in the case of a very small sample from a very small population. The population is the scores of 10 students on an exam:

Student	0	1	2	3	4	5	6	7	8	9
Score	82	62	80	58	72	73	65	66	74	62

The parameter of interest is the mean score μ in this population. The sample is an SRS of size $n = 4$ drawn from the population.

 (a) Find the mean of the 10 scores in the population. This is the population mean μ.

(b) Use the **SAMPLE** command to draw an SRS of size 4 from this population. Write the four scores in your sample and calculate the mean $\bar{x}$ of the sample scores. This statistic is an estimate of μ.

(c) Repeat this process 10 times. Make a histogram of the 10 values of $\bar{x}$. You are constructing the sampling distribution of $\bar{x}$. Is the center of your histogram close to μ?

4.47 Table 1.9 and TA01-09.MTW give the survival times of 72 guinea pigs in a medical experiment. Consider these 72 animals to be the population of interest.

(a) Make a histogram of the 72 survival times. This is the population distribution. It is strongly skewed to the right.

(b) Find the mean of the 72 survival times. This is the population mean μ. Mark μ on the x axis of your histogram.

(c) Use the **SAMPLE** command to choose an SRS of size $n = 12$. What is the mean survival time $\bar{x}$ for your sample? Mark the value of $\bar{x}$ with a point on the axis of your histogram from (a).

(d) Choose four more SRSs of size 12. Find $\bar{x}$ for each sample and mark the values on the axis of your histogram from (a). Would you be surprised if all five $\bar{x}$'s fell on the same side of μ? Why?

(e) If you chose a large number of SRSs of size 12 from this population and made a histogram of the $\bar{x}$ values, where would you expect the center of this sampling distribution to lie?

4.69 The rate of return on a stock varies from month to month. We can use a control chart to see whether the pattern of variation is stable over time or whether there are periods during which the stock was unusually volatile by comparison with its own long-run pattern. In EX04-69.MTW, we have data on the monthly returns (in percent) on Wal-Mart common stock during its early years of rapid growth, 1973 through 1991. Consider these 228 observations as 38 subgroups of 6 consecutive months each.

(a) Make an **XBACHART** and **SCHART** for the monthly return on Wal-Mart common stock.

(b) Based on your examination of the s chart, explain why the control limits for your $\bar{x}$ chart are so wide as to be of little use.

Sometimes we want to make a control chart for a single measurement x at each time period. Control charts for individual observations can be produced using the Minitab command **ICAHRT** or by selecting **Stat ➤ Control Charts ➤ Individu-**

als from the menu. Control charts for individual measurement are just $\bar{x}$ charts with the sample size $n = 1$. We cannot estimate the short-term process standard deviation σ from the individual samples because a sample of size 1 has no variation. To compensate for this, it is common to use 2σ rather than 3σ, that is, control limits at $\mu \pm 2\sigma$. The next two exercises illustrate control charts for individual measurements.

4.79 Professor Moore swims 2000 yards regularly. Here and in EX04-79.MTW are his times (in minutes) for 23 sessions in the pool:

34.12	35.72	34.72	34.05	34.13	35.72	36.17	35.57
35.37	35.57	35.43	36.05	34.85	34.70	34.75	33.93
34.60	34.00	34.35	35.62	35.68	35.28	35.97	

(a) Use the **ICAHRT** command and the **SLIMITS** subcommand to make a control chart for the 23 times with center line at $\bar{x}$ and control limits at $\bar{x} \pm 2s$.

(b) Are the professor's swimming times in control? If not, describe the nature of any lack of control.

4.80 Professor Moore records the time he takes to drive to the college each morning. Here and in EX04-80.MTW are the times (in minutes) for 42 consecutive weekdays, with the dates in order along the rows:

8.25	7.83	8.30	8.42	8.50	8.67	8.17	9.00	9.00	8.17	7.92
9.00	8.50	9.00	7.75	7.92	8.00	8.08	8.42	8.75	8.08	9.75
8.33	7.83	7.92	8.58	7.83	8.42	7.75	7.42	6.75	7.42	8.50
8.67	10.17	8.75	8.58	8.67	9.17	9.08	8.83	8.67		

He also noted unusual occurrences on his record sheet: on October 27, a truck backing into a loading dock delayed him, and on December 5, ice on the windshield forced him to stop and clear the glass.

(a) Use the **ICAHRT** command and the **SLIMITS** subcommand to make a control chart for the driving times with center line at $\bar{x}$ and control limits at $\bar{x} \pm 2s$. (The standard deviation s of all observations includes any long-term variation, so these limits are a bit crude.)

(b) Comment on the control of the process. Can you suggest explanations for individual points that are out of control? Is there any indication of an upward or downward trend in driving time?

Chapter 5
Probability Theory

Commands to be covered in this chapter:

```
RANDOM K observations into C...C
PDF for values in E...E [put results in E...E]
CDF for values in E...E [put results in E...E]
```

The RANDOM Command

The **RANDOM** command is used to generate random numbers. The format for the command is

```
RANDOM K observations into C...C
```

where K is the number of random numbers to be generated and C...C are the column(s) in which the random numbers are to be stored. The command can also be used by selecting **Calc ➤ Random Data** and a probability distribution from the menu.

In example 5.8 of BPS, an engineer chooses a sample of 10 switches from a shipment. Suppose that 10% of the switches in the shipment are bad. The engineer will count the number X of bad switches. In Chapter 4 we learned to generate random numbers for this situation with the **BERNOULLI** subcommand or by selecting **Calc ➤ Random Data ➤ Bernoulli** from the menu. Here we generate a sequence of 10 1's and 0's to represent the bad and good switches.

```
MTB > Random 10 c1;
SUBC> Bernoulli .1.
MTB > print c1
```

Data Display

```
C1
    0    0    0    1    0    0    0    0    0    0
```

If we are interested only in the number of bad switches, we could just generate the number X using the **RANDOM** command with the **BINOMIAL** subcommand. The subcommand format is

BINOMIAL n = K, p = K

where n is the number of trials and p is the probability of success on each trial. In this example, $n = 10$ and $p = 0.1$. Alternatively, the command can be used by selecting **Calc ➤ Random Data ➤ Binomial** from the menu.

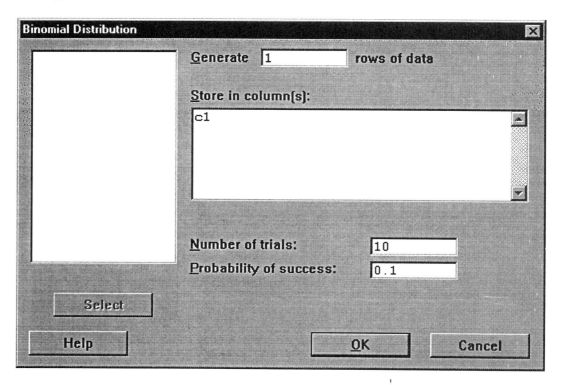

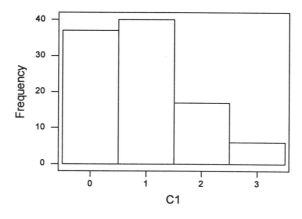

Instead of generating only one value for X, we can use the **RANDOM** command to simulate a large number of repetitions of the sample. If we change the dialog box to generate 100 rows of data, we could examine how X varies from sample to sample. The histogram shows that it is most likely that X would be equal to 1, but that in our 100 repetitions, X is also equal to 0, 2, and 3.

The PDF Command

The **PDF** (probability distribution function) command has a special capability when used with the subcommand **BINOMIAL**. If no arguments are specified on the **PDF** command line, then the probability distribution of that binomial distribution will be printed. Below we obtain the probability of each possible outcome for a binomial distribution with $n = 10$ and $p = 0.1$.

```
MTB > pdf;
SUBC> binomial 10 .1.
```

Probability Density Function

```
Binomial with n = 10 and p = 0.100000

        x         P( X = x)
        0           0.3487
        1           0.3874
        2           0.1937
        3           0.0574
        4           0.0112
        5           0.0015
        6           0.0001
        7           0.0000
```

Note that for K = 7, $P(X = x)$ is equal to 0 (rounded to 4 decimal places). For K = 8, 9, and 10, $P(X = x)$ is also equal to 0, so these rows are not printed in the table of probabilities for the binomial.

The CDF Command

In addition to the **PDF** command, the **CDF** (cumulative distribution function) command also can be used to print a table of probabilities for the binomial distribution. The difference is that this command gives cumulative probabilities, that is, the probability that X is less than or equal to a value. Below we use Minitab to obtain the cumulative probabilities for a binomial with $n = 10$ and $p = 0.1$.

```
MTB > cdf;
SUBC> bino 10 .1.
```

104   Chapter 5

Cumulative Distribution Function

```
Binomial with n = 10 and p = 0.100000

        x        P( X <= x)
        0          0.3487
        1          0.7361
        2          0.9298
        3          0.9872
        4          0.9984
        5          0.9999
        6          1.0000
```

The probability that X is less than or equal to 7, 8, 9, or 10 is equal to 1, so these rows are not printed in the above table.

Selecting **Calc ➤ Probability Distributions ➤ Binomial** from the menu also allows you to calculate probabilities and cumulative probabilities for a binomial distribution. The results can be stored in a Minitab worksheet if you specify a storage column for the generated values. Below, we show how the cumulative probabilities can be obtained. The input column is required and contains the numbers (0,1,2,...,10) for which we want the cumulative probability. The number of trials and probability of success are also entered in the dialog box.

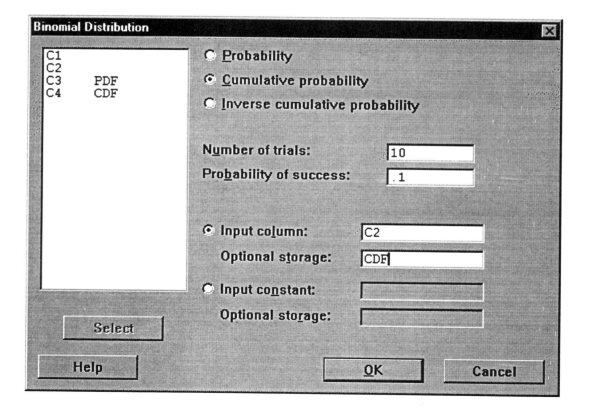

Suppose that an opinion poll (as described in BPS Example 5.12) asks 2500 adults whether they agree or disagree that "I like buying new clothes, but shopping is often frustrating and time-consuming." Suppose also that 60% percent of all adult U.S. residents would say "Agree".

We can use the **RANDOM** command to simulate possible outcomes from this survey. The **RANDOM** command and the **BINOMIAL** subcommand are used below to generate the number of successes from 2500 independent trials where the answer "Yes," is considered a success.

```
MTB > random 1000 c1;
SUBC> binomial 2500 .6.
MTB > name c1 'Count X'
```

We can illustrate the distribution of the count of X using the **HISTOGRAM** command. An alternative is the **%DESCRIBE** macro. The **%DESCRIBE** macro generates output in a Graph window. The Graph window includes a histogram with a normal curve, a boxplot, confidence intervals for the mean and median, and a table of statistics.

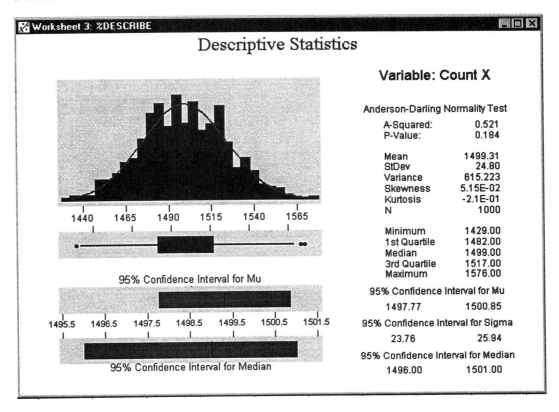

As the window shows, the normal distribution approximated the binomial well. The sample values for the mean and standard deviation are close to the theoretical values of

$$\mu = np = (2500)(.6) = 1500$$
$$\sigma = \sqrt{np(1-p)} = \sqrt{2500(.6)(.4)} = 24.49.$$

The values are easily calculated using Minitab's calculator (choose **Calc ➤ Calculator** from the menu) or the **LET** command.

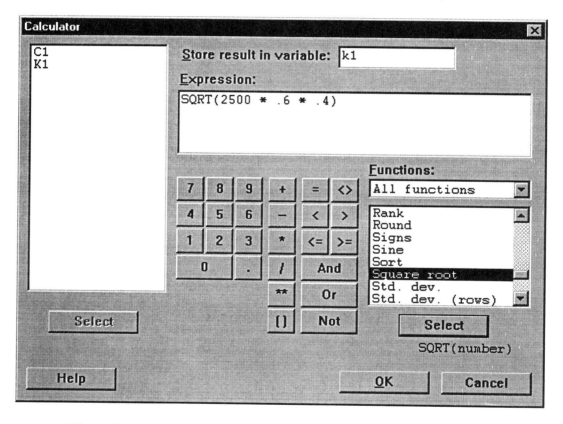

When the normal approximation to the binomial applies ($np \geq 10$ and $n(1-p) \geq 10$), the **CDF** command can be used to approximate binomial probabilities. The command calculates the probability that an observation is less than or equal to each value in E. The command format is as follows:

```
CDF for values in E [store results in E]
```

This command can be selected from the menu by choosing **Calc ➤ Probability Distributions ➤ Normal.** Below, we approximate the probability that at least 1520 of the people in the sample find shopping frustrating, when $n = 2500$ and $p = 0.6$. We act as though the count X has the $N(1500, 24.49)$ distribution. The **CDF** command calculates the probability that an observation is less than or equal to a value, so we then subtract the result from one.

```
MTB > cdf 1520;
SUBC> normal 1500 24.49.
```

Cumulative Distribution Function

```
Normal with mean = 1500.00 and standard deviation = 24.4900

         x       P( X <= x)
   1.52E+03        0.7929

MTB > let k1 = 1-.7929
MTB > print k1
```

Data Display

```
K1     0.207100
```

Therefore, the probability we want is approximated to be 0.2071. The same result could be obtained by selecting **Calc ➤ Probability Distributions ➤ Normal** from the menu and filling in the dialog box as shown.

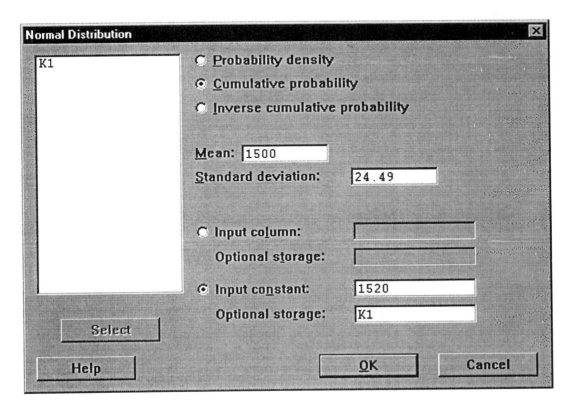

EXERCISES

5.33 Among employed women, 25% have never been married. Select 10 employed women using the **RANDOM** command.
    (a) The number in your sample who have never been married has a binomial distribution. What are n and p?
    (b) Use the **PDF** command with the **BINOMIAL** subcommand to find the probability that exactly 2 of the 10 women in your sample have never been married.
    (c) Use the **CDF** command with the **BINOMIAL** subcommand to find the probability that 2 or fewer have never been married.
    (d) Use Minitab's calculator or the **LET** command to find the mean number of women in such samples who have never been married. What is the standard deviation?

5.34 In a test for ESP (extrasensory perception), a subject is told that cards the experimenter can see but he cannot contain either a star, a circle, a wave, or a square. As the experimenter looks at each of 20 cards in turn, the subject names the shape on the card. A subject who is just guessing has probability 0.25 of guessing correctly on each card.
    (a) The count of correct guesses in 20 cards has a binomial distribution. What are n and p?
    (b) Use Minitab's calculator or the **LET** command to find the mean number of correct guesses in many repetitions.
    (c) Use the **PDF** command with the **BINOMIAL** subcommand to find the probability of exactly 5 correct guesses.

5.35 A believer in the "random walk" theory of stock markets thinks that an index of stock prices has probability 0.65 of increasing in any year. Moreover, the change in the index in any given year is not influenced by whether it rose or fell in earlier years. Let X be the number of years among the next 5 years in which the index rises.
    (a) X has a binomial distribution. What are n and p?
    (b) Use the **RANDOM** command to simulate 100 replications of X.
    (c) Make a **HISTOGRAM** of your simulated values of X and use the **DESCRIBE** command find the mean and standard deviation of this distribution.

5.37. Here is a simple probability model for multiple-choice tests. Suppose that each student has probability p of correctly answering a question chosen at random from a universe of possible questions. (A strong student has a higher p than a weak student.) Answers to different questions are independent. Jodi is a good student for whom $p = 0.75$.
   (a) Use Minitab's calculator or the **LET** command to find the mean and standard deviation for the number of questions Jodi will answer correctly. Use the normal approximation and the **CDF** command to find the probability that Jodi scores 70% or lower on a 100-question test.
   (b) If the test contains 250 questions, what is the probability that Jodi will score 70% or lower?

5.54 Leakage from underground gasoline tanks at service stations can damage the environment. It is estimated that 25% of these tanks leak. You examine 15 tanks chosen at random, independently of each other.
   (a) Use the **RANDOM** command to simulate 100 replications of this larger study. Examine the shape of the distribution. Is it reasonable to use the normal approximation?
   (b) Use the **CDF** command with the **BINOMIAL** subcommand to find the probability that 10 or more of the 15 tanks leak.
   (c) Now you do a larger study, examining a random sample of 1000 tanks nationally. Use the **RANDOM** command to simulate 100 replications of this larger study. Examine the shape of the distribution. Is it reasonable to use the normal approximation? What is the probability that at least 275 of these tanks are leaking?

5.55 An opinion poll asks an SRS of 500 adults whether they favor giving parents of school-age children vouchers that can be exchanged for education at any public or private school of their choice. Each school would be paid by the government on the basis of how many vouchers it collected. Suppose that in fact 45% of the population favor this idea. Use the normal approximation and the **CDF** command to find the probability that more than half of the sample are in favor.

5.57 While he was a prisoner of the Germans during World War II, John Kerrich tossed a coin 10,000 times. He got 5067 heads. Take Kerrich's tosses to be an SRS from the population of all possible tosses of his coin. If the coin is perfectly balanced, $p = 0.5$. Is there reason to think that Kerrich's coin gave too many heads to be balanced?

(a) Use Minitab's calculator or the **LET** command to find the mean and standard deviation for the number of heads Kerrich should have gotten from 10,000 coin tosses.

(b) Use the normal approximation and the **CDF** command to find the probability that a balanced coin would give 5067 or more heads in 10,000 tosses. What do you conclude?

Chapter 6
Introduction to Inference

Commands to be covered in this chapter:

```
ZINTERVAL [K% confidence], sigma = K for C...C
ZTEST [of m = K], assuming sigma = K for C...C
INVCDF for values in E [put into E]
```

The ZINTERVAL Command

The **ZINTERVAL** command calculates a normal theory confidence interval for the mean, with σ known, separately on each column. The command format follows.

```
ZINTERVAL [K% confidence], sigma = K for C...C
```

This interval goes from $\bar{x} - z^*\left(\sigma/\sqrt{n}\right)$ to $\bar{x} + z^*\left(\sigma/\sqrt{n}\right)$ where $\bar{x}$ is the mean of the data, n is the sample size, and z^* is the value from the normal table corresponding to K percent confidence. If K is not specified, K = 95 is used. If the value of K for the confidence level is less than one, it is assumed to be a confidence coefficient and not a percentage, and is multiplied by 100. Confidence intervals can also be calculated by selecting **Stat ➤ Basic Statistics ➤ 1-Sample Z** from the menu. In the dialog box, enter the column containing the samples in the Variables box, click on Confidence interval and specify a Level. Enter a value for σ in the Sigma box and click OK.

Example 6.4 of BPS describes a manufacturer of pharmaceutical products. The laboratory verifies the concentration of active ingredients by analyzing each specimen 3 times. The standard deviation of this distribution is known to be $\sigma = 0.0068$ grams per liter. Three analyses of one specimen give concentrations

$$0.8403 \quad 0.8363 \quad 0.8447$$

We want a 99% confidence interval for the true concentration μ. We enter the data into a Minitab worksheet and then calculate the confidence interval.

112   Chapter 6

```
MTB > set c1
DATA> .8403 .8363 .8447
DATA> end
MTB > name c1 'concentration'
```

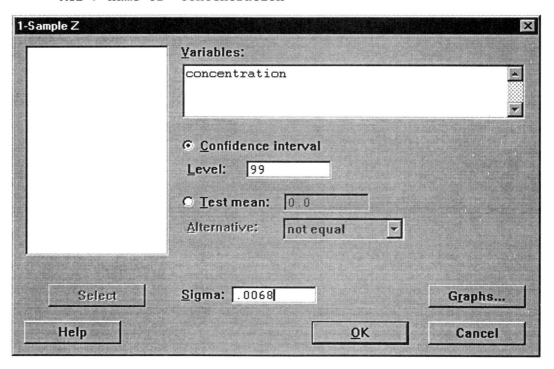

```
MTB > ZInterval 99 .0068 'concentration'.
```

Z Confidence Intervals

```
The assumed sigma = 0.00680

Variable   N     Mean    StDev   SE Mean       99.0 % CI
concentr   3   0.84043  0.00420  0.00393  ( 0.83032, 0.85055)
```

Note that we specified a 99% interval by letting K = 99. Minitab calculates a 95% confidence interval when we do not specify a value for K. The values for Mean and StDev listed with the confidence intervals are the same as those that would be obtained using the **DESCRIBE** command. The value given for SE Mean is calculated with the known value of σ as follows.

$$\frac{\sigma}{\sqrt{n}} = \frac{.0068}{\sqrt{40}} = .00108.$$

The ZTEST Command

The `ZTEST` command performs a normal theory test with σ known, separately on each column. If K is not specified, $\mu = 0$ is used. The command format is

```
ZTEST [of m = K], assuming sigma = K for C...C
```

If no subcommand is given, a two-sided test is done. Otherwise, a subcommand specifies the alternative hypothesis with the following form.

```
ALTERNATIVE = K
```

If `ALT = -1`, then $\mu < K$ is used. If `ALT = 1`, $\mu > K$ is used. Note that the `ALTERNATIVE` subcommand requires only three letters. If we wish to test the hypothesis that $\mu = 0.86$ against the alternative $\mu \neq 0.86$ for the data given above, we do not need to use a subcommand, as illustrated below.

```
MTB > ztest 86 .0068 'concentration'
```

Z-Test

```
Test of mu = 86.0000 vs mu not = 86.0000
The assumed sigma = 0.00680

Variable    N      Mean     StDev    SE Mean         Z         P
concentr    3    0.8404    0.0042     0.0039-21691.28    0.0000
```

If instead we wish to test the hypothesis that $\mu = 0.86$ against the alternative $\mu < 0.86$ for the data given above, we use the `ALTERNATIVE` subcommand with $K = -1$. This is illustrated below.

```
MTB > ztest .86 .0068 c1;
SUBC> alternative -1.
```

Z-Test

```
Test of mu = 86.0000 vs mu < 86.0000
The assumed sigma = 0.00680

Variable    N      Mean     StDev    SE Mean         Z         P
concentr    3    0.8404    0.0042     0.0039-21691.28    0.0000
```

The *P*-value given is smaller for the one-sided test. In fact, it is equal to half the *P*-value computed for the two-sided test. In both tests, the value is very small and the null hypothesis should be rejected.

114   *Chapter 6*

As with confidence intervals, we can **select Stat ➤ Basic Statistics ➤ 1-Variable Z** to do a hypothesis test. The appropriate alternative is selected in the dialog box.

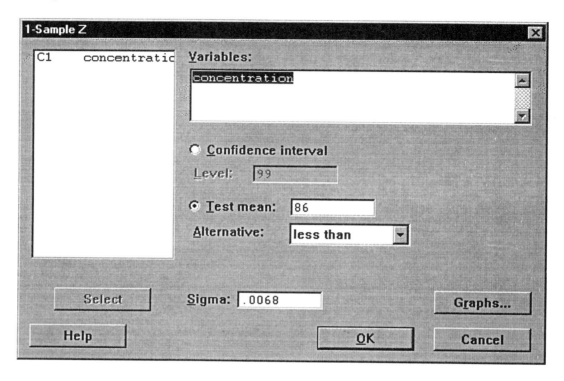

The INVCDF Command

The **INVCDF** command can be used to find the critical value that would be required to reject the null hypothesis at a particular significance level. The **INVCDF**, which was introduced in Chapter 1, calculates a value associated with an area. We can find the value of z that has a specific area below it. If no subcommand is specified, then the distribution is assumed to be normal with $\mu = 0$ and $\sigma = 1$. The **INVCDF** command has the following format.

```
INVCDF for values in E [store results in E]
```

Suppose that we wish to use a 1% significance criterion for the example above. Since the test is one-sided, the probability in the tail must be 0.01 or less, so we find the value of z that has area 0.01 below it.

```
MTB > invc .01
```

Inverse Cumulative Distribution Function

```
Normal with mean = 0 and standard deviation = 1.00000

P( X <= x)              x
   0.0100          -2.3263
```

Therefore, we would reject the null hypothesis at the 1% level of significance if $z < -2.3263$.

EXERCISES

6.5   Here and in EX06-05.MTW are the IQ test scores of 31 seventh-grade girls in a Midwest school district:

114	100	104	89	102	91	114	114	103	105	108
130	120	132	111	128	118	119	86	72	111	
103	74	112	107	103	98	96	112	112	93	

(a)   We expect the distribution of IQ scores to be close to normal. Make a **STEM-AND-LEAF** display of the distribution of these 31 scores. Does your plot show outliers, clear skewness, or other nonnormal features?

(b)   Treat the 31 girls as an SRS of all seventh-grade girls in the school district. Suppose that the standard deviation of IQ scores in this population is known to be $\sigma = 15$. Use the **ZINTERVAL** command to give a 99% confidence interval for the mean score in the population.

(c)   In fact, the scores are those of all seventh-grade girls in one of the several schools in the district. Explain carefully why your confidence interval from (b) cannot be trusted.

6.7   Example 6.4 gives confidence intervals for the concentration μ based on three measurements (0.8403, 0.8363, and 0.8447) with $\bar{x} = 0.8404$ and $\sigma = 0.0068$. The 99% confidence interval is 0.8303 to 0.8505.

(a)   Enter the three measurements (0.8403, 0.8363, and 0.8447) into a Minitab worksheet. Use the **ZINTERVAL** command to find the 80% confidence interval for μ.

(b)   Use the **ZINTERVAL** command to find the 99.9% confidence interval for μ.

116   Chapter 6

(c) How does increasing the confidence level affect the length of the confidence interval?

6.19 Biologists studying the healing of skin wounds measured the rate at which new cells closed a razor cut made in the skin of an anesthetized newt. Here are data from 18 newts, measured in micrometers (millionths of a meter) per hour:

| 29 | 27 | 34 | 40 | 22 | 28 | 14 | 35 | 26 |
| 35 | 12 | 30 | 23 | 18 | 11 | 22 | 23 | 33 |

(a) Make a **STEM-AND-LEAF** display of the healing rates. It is difficult to assess normality from 18 observations, but look for outliers or extreme skewness. What do you find?

(b) Scientists usually assume that animal subjects are SRSs from their species or genetic type. Treat these newts as an SRS and suppose you know that the standard deviation of healing rates for this species of newt is 8 micrometers per hour. Use the **ZINTERVAL** command to give a 90% confidence interval for the mean healing rate for the species.

(c) A friend who knows almost no statistics follows the formula $\bar{x} \pm 2\sigma/\sqrt{n}$ in a biology lab manual to get 95% confidence interval for the mean. Is her interval wider or narrower than yours? Explain to her why it makes sense that higher confidence changes the length of the interval.

6.20 Below and in EX06-20.MTW are measurements (in millimeters) of a critical dimension on a sample of auto engine crankshafts.

224.120	224.001	224.017	223.982	223.989	223.961
223.960	224.089	223.987	223.976	223.902	223.980
224.098	224.057	223.913	223.999		

The data come from a production process that is known to have standard deviation $\sigma = 0.060$ mm. The process mean is supposed to be $\mu = 224$ mm but can drift away from this target during production.

(a) We expect the distribution of the dimension to be close to normal. Make a stemplot or histogram of these data and describe the shape of the distribution.

(b) Use the **ZINTERVAL** command to compute a 95% confidence interval for the process mean at the time these crankshafts were produced.

6.35 In Exercise 6.20 and in EX06-20.MTW are measurements (in millimeters) of a critical dimension on a sample of automobile engine crankshafts. The manufacturing process is known to vary normally with standard deviation $\sigma = 0.060$ mm. The process mean is supposed to be 224 mm. Do these data give evidence that the process mean is not equal to the target value 224 mm? Use the **ZTEST** command to find the P-value for the following hypothesis test.

$$H_0 : \mu = 224$$
$$H_a : \mu \neq 224$$

6.36 Bottles of a popular cola are supposed to contain 300 milliliters (ml) of cola. There is some variation from bottle to bottle because the filling machinery is not perfectly precise. The distribution of the contents is normal with standard deviation $\sigma = 3$ ml. An inspector who suspects that the bottler is underfilling measures the contents of six bottles. The results are

299.4 297.7 301.0 298.9 300.2 297.0

Is this convincing evidence that the mean content of cola bottles is less than the advertised 300 ml?

(a) State the hypotheses that you will test.
(b) Use the **ZTEST** command to calculate the test statistic and find the P-value.
(c) State your conclusion.

6.37 A random number generator is supposed to produce random numbers that are uniformly distributed in the interval from 0 to 1. If this is true, the numbers generated come from a population with $\mu = 0.5$ and $\sigma = 0.2887$. Simulate 100 numbers using the **RANDOM** command with the **UNIFORM** subcommand. We want to test

$$H_0 : \mu = 0.5$$
$$H_a : \mu \neq 0.5$$

(a) Use the **ZTEST** command to find the P-value for this test.
(b) Is the result significant at the 5% level ($\alpha = 0.05$)?
(c) Is the result significant at the 1% level ($\alpha = 0.01$)?

118   Chapter 6

6.39   In exercise 6.5 and in EX06-05.MTW are the IQ test scores of 31 seventh-grade girls in a Midwest school district. Treat the 31 girls as an SRS of all seventh-grade girls in the school district. Suppose that the standard deviation of IQ scores in this population is known to be $\sigma = 15$.

   (a)   Use the **ZINTERVAL** command to give a 95% confidence interval for the mean IQ score μ in the population.

   (b)   Is there significant evidence at the 5% level that the mean IQ score in the population differs from 100? State hypotheses and use your confidence interval to answer the question without more calculations.

6.74   Sulfur compounds cause "off-odors" in wine, so winemakers want to know the odor threshold, the lowest concentration of a compound that the human nose can detect. The odor threshold for dimethyl sulfide (DMS) in trained wine tasters is about 25 micrograms per liter of wine (μg/l). The untrained noses of consumers may be less sensitive, however. Here are the DMS odor thresholds for 10 untrained students:

| 31 | 31 | 43 | 36 | 23 | 34 | 32 | 30 | 20 | 24 |

Assume that the standard deviation of the odor threshold for untrained noses is known to be $\sigma = 7$ μg/l.

   (a)   Make a **STEM-AND-LEAF** display to verify that the distribution is roughly symmetric with no outliers. (More data confirm that there are no systematic departures from normality.)

   (b)   Use the **ZINTERVAL** command to give a 95% confidence interval for the mean DMS odor threshold among all students.

   (c)   Are you convinced that the mean odor threshold for students is higher than the published threshold, 25 μg/l? Use the **ZTEST** command to carry out a significance test to justify your answer.

6.77   U.S. Treasury bills are safe investments, but how much do they pay investors? Here and in EX 06-77.MTW are data on the total returns (in percent) on Treasury bills for the years 1970 to 1996:

Year	1970	1971	1972	1973	1974	1975	1976	1977	1978
Return	6.45	4.37	4.17	7.20	8.00	5.89	5.06	5.43	7.46
Year	1979	1980	1981	1982	1983	1984	1985	1986	1987
Return	10.56	12.18	14.71	10.84	8.98	9.89	7.65	6.10	5.89
Year	1988	1989	1990	1991	1992	1993	1994	1995	1996
Return	6.95	8.43	7.72	5.46	3.50	3.04	4.37	5.60	5.13

(a) Make a **HISTOGRAM** of these data, using bars 2 percentage points wide. What kind of deviation from normality do you see? Thanks to the central limit theorem, we can nonetheless treat $\bar{x}$ as approximately normal.

(b) Suppose that we can regard these 27 years' results as a random sample of returns on Treasury bills. Use the **ZINTERVAL** command to give a 90% confidence interval for the long-term mean return. (Assume you know that the standard deviation of all returns is $\sigma = 2.75\%$.)

(c) The rate of inflation during these years averaged about 5.5%. Are you convinced that Treasury bills have a mean return higher than 5.5%? State hypotheses and use the **ZTEST** command. Give a test statistic and a *P*-value.

(d) Make a **TSPLOT** of the data. There are strong up-and-down cycles in the returns, which follow cycles of interest rates in the economy. The time plot makes it clear that returns in successive years are strongly correlated, so it is not proper to treat these data as an SRS. You should always check for such *serial correlation* in data collected over time.

Chapter 7
Inference for Distributions

Commands to be covered in this chapter:

```
TINTERVAL [K% confidence] for data in C...C
TTEST [of m = K] on data in C...C
PAIR C C
TWOSAMPLE test and CI [K% conf] samples in C C
TWOT test with [K% conf] data in C, groups in C
```

The TINTERVAL Command

The **TINTERVAL** command calculates a *t* confidence interval for the mean. The confidence interval is calculated separately for each column specified in the following format.

```
TINTERVAL [K% confidence] for data in C...C
```

If the percent confidence is not specified, a 95% confidence interval is calculated.

Example 7.1 of BPS describes a study of insect metabolism. Researchers fed cockroaches measured amounts of a sugar solution. After 2, 5, and 10 hours, they dissected some of the cockroaches and measured the amount of sugar in various tissues. Five roaches fed the sugar D-glucose and dissected after 10 hours had the following amounts (in micrograms) of D-glucose in their hindguts:

$$55.95 \quad 68.24 \quad 52.73 \quad 21.50 \quad 23.78$$

Using the **TINTERVAL** command, we give a 95% confidence interval for the mean amount of D-glucose in cockroach hindguts under these conditions.

    MTB > tint 'sugar'

T Confidence Intervals

```
Variable    N     Mean    StDev   SE Mean         95.0 % CI
sugar       5    44.44    20.74      9.28  (   18.69,    70.19)
```

If we use the **DESCRIBE** command to calculate the Mean, StDev, and SE Mean, we note that the results are the same as above.

    MTB > desc 'sugar'

Descriptive Statistics

```
Variable          N     Mean   Median    Mean    StDev   SE Mean
sugar             5    44.44    52.73   44.44    20.74      9.28

Variable    Minimum  Maximum       Q1       Q3
sugar         21.50    68.24    22.64    62.09
```

The **TTEST** Command

The **TTEST** command performs a separate *t* test for each column specified in the following format.

    TTEST [of mu = K] on data in C...C

By default, Minitab does a two-sided test with the test mean $\mu = 0$. Other values for μ can be specified. To do one-sided tests, use the **ALTERNATIVE** subcommand. As with the **ZTEST** command, the subcommand format is

    ALTERNATIVE = K

If ALT = -1, then $\mu < K$ is used. If ALT = +1, $\mu > K$ is used. Below we illustrate the **TTEST** command by testing whether or not a new cola lost sweetness during storage as described in Example 7.2 of BPS. Trained tasters rate the sweetness before and after storage. Here are the sweetness losses found by 10 tasters for one new cola recipe.

       2.0   0.4   0.7   2.0   −0.4   2.2   −1.3   1.2   1.1   2.3

We use Minitab to test

$$H_0: \mu = 0$$
$$H_0: \mu > 0$$

```
MTB > ttest c1;
SUBC> alt +1.
```

T-Test of the Mean

```
Test of mu = 0.000 vs mu > 0.000

Variable     N      Mean    StDev   SE Mean      T        P
losses      10     1.020    1.196     0.378    2.70    0.012
```

The *P*-value is reported as 0.012. There is quite strong evidence for a loss of sweetness. We can safely reject H_0.

Both the **TINTERVAL** and the **TTEST** commands can be used from the menu by selecting **Stat ➤ Basic Statistics ➤ 1-Sample t** and filling out the dialog box. The dialog box for 1-Sample t is similar to the dialog box for 1-Sample Z, except that it is not required to enter a value for Sigma. The dialog box includes a button for Graphs. This is used to check the data graphically for outliers or strong skewness that might threaten the validity of the *t* procedures.

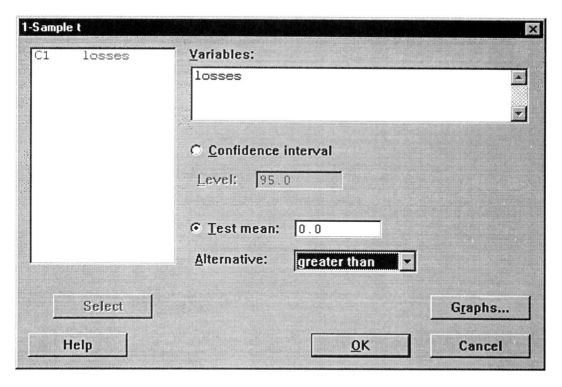

The `INVCDF` Command

To determine the critical value of t that would be required to reject the null hypothesis, we can use the `INVCDF` command with the `T` subcommand. The subcommand format for the t distribution with K degrees of freedom is

```
T with degrees of freedom = K
```

Since the above test was two-sided, we would reject at the 5% level of significance as long as the probability of being in each tail is less than 2.5%. Below, we use Minitab to find the value of t with area 0.025 below it. We specify that there are 4 degrees of freedom since $n = 5$.

```
MTB > invcdf .025;
SUBC> t 4.
```

Inverse Cumulative Distribution Function

```
Student's t distribution with 4 DF

P( X <= x)            x
    0.0250      -2.7764
```

Therefore, we would reject H_0 for any value of t that is less than -2.7764. Since the test is two-sided and the t distribution is symmetric, we would also reject H_0 for any value of t that is larger than 2.7764.

Matched Pairs

In a matched pairs study, subjects are matched in pairs and the outcomes are compared within each matched pair. In example 7.3 of BPS, subjects worked a paper-and-pencil maze while wearing masks. Each mask was either unscented or carried a floral scent. The response variable is their mean time on three trials. Each subject worked the maze with both types of mask. Table 7.1 and TA07-01.MTW give the subjects' times.

```
MTB > info
```

Information on the Worksheet

```
Column   Count  Name
C1          21  Order
C2          21  Unscented
C3          21  Scented
```

To determine whether the subjects' times improved, we will test

$$H_0: \mu = 0$$
$$H_a: \mu > 0$$

where we are considering the improvement from the floral scent.

```
MTB > pair c2 c3;
SUBC> alt +1.
```

Paired T-Test and Confidence Interval

```
Paired T for Unscented - Scented

              N      Mean     StDev    SE Mean
Unscente     21     50.01     14.36      3.13
Scented      21     49.06     13.39      2.92
Difference   21      0.96     12.55      2.74

95% CI for mean difference: (-4.76, 6.67)
T-Test of mean difference = 0 (vs > 0):
           T-Value = 0.35   P-Value = 0.365
```

The large value given for the *P*-value means that the data do not support the claim that floral scents improve performance. A 95% confidence interval for the mean improvement was computed. Note that the computed interval includes 0.

Selecting **Stat** ➤ **Basic Statistics** ➤ **Paired t** from the menu is another way to do a *t* test for matched pairs. The null hypothesis test value and alternative hypothesis are specified by clicking the Options button and filling out the Options subdialog box.

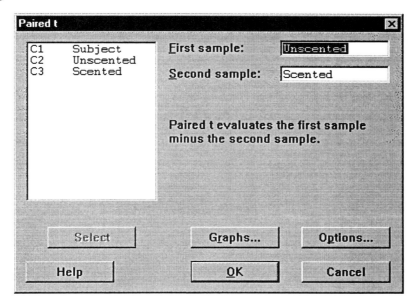

Two-Sample t: Confidence Interval and Test

Select **Stat ➤ Basic Statistics ➤ 2-Sample-t** from the menu to perform a hypothesis test and compute a confidence interval of the difference between two population means. The data can be entered in one of two ways. Either both samples are in a single column with another column of subscripts to identify the population group, or each sample can be in a separate column.

Example 7.7 of BPS fits the two-sample setting. A researcher buried polyester strips in the soil to see how quickly they decay. Five of the strips, chosen at random, were dug up after two weeks. Another five were dug up after 16 weeks. The breaking strength (in pounds) of all 10 strips was measured and entered into EG07-07.

2 weeks	118	126	126	120	129
16 weeks	124	98	110	140	110

Since the samples were in one column, we checked "Samples in one column" in the dialog box and enter the appropriate columns under "Samples:" and "Subscripts:". Since the hypothesis are

$$H_0: \mu_1 = \mu_1$$
$$H_a: \mu_1 > \mu_1$$

we select the greater than alternative.

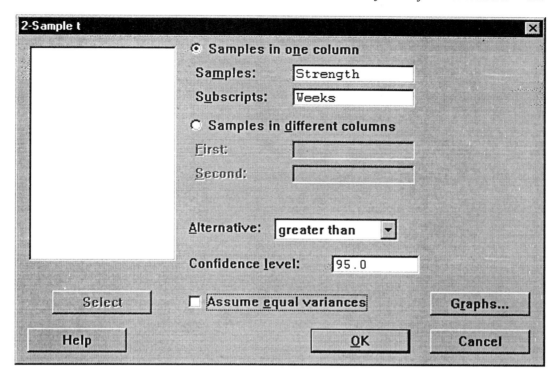

Two Sample T-Test and Confidence Interval

```
Two sample T for Strength

Weeks      N      Mean    StDev    SE Mean
  2        5    123.80     4.60      2.1
 16        5    116.4     16.1       7.2

95% CI for mu ( 2) - mu (16): ( -13.4,   28.2)
T-Test mu ( 2) = mu (16) (vs >): T = 0.99   P = 0.19   DF = 4
```

The *P*-value was calculated to be 0.19. The experiment did not find convincing evidence that polyester decays more in 16 weeks than in two weeks. The output also shows that the 95% confidence interval for $\mu_1 - \mu_2$ is (−13.4, 28.2). The Graphs button could be selected to produce a graph illustrating the differences between the two groups. We selected boxplots in the Graphs subdialog box.

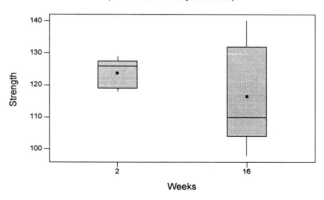

Boxplots of Strength by Weeks
(means are indicated by solid circles)

Two Minitab session commands are available to do two-sample t procedures. These are the **TWOSAMPLE** command and the **TWOT** command. The commands differ only in the way the input data are arranged. The **TWOSAMPLE** command performs a two (independent) sample t test and confidence interval with the data for each sample in separate columns. The command format is given below.

```
TWOSAMPLE test and CI [K% conf] samples in C C
```

The **TWOT** command also performs a two (independent) sample t test and confidence interval. This command is used when the data for both samples are in one column. A second column contains sample subscripts. The command format follows.

```
TWOT test with [K% conf] data in C, groups in C
```

Unless the **POOLED** subcommand is used, the **TWOSAMPLE** and the **TWOT** commands do not assume that the populations have equal variances, so the test statistic is

$$t = \frac{\overline{X}_1 - \overline{X}_2}{\sqrt{\frac{s_1^2}{n_1} + \frac{s_2^2}{n_2}}}$$

This statistic has approximately a t distribution with degrees of freedom given by:

$$df = \frac{\left(\frac{s_1^2}{n_1} + \frac{s_2^2}{n_2}\right)^2}{\frac{1}{n_1-1}\left(\frac{s_1^2}{n_1}\right)^2 + \frac{1}{n_2-1}\left(\frac{s_2^2}{n_2}\right)^2}$$

Minitab truncates the number to an integer, if necessary. The same calculation is used when the Two-Sample t procedure is used from the menu provided that Assume Equal Variances is not checked.

If the **POOLED** subcommand is used or Assume Equal Variances is checked in the Two-Sample t dialog box, Minitab uses a pooled procedure, which assumes the two populations have equal variances and "pools" the two sample variances to estimate the common population variance. The test statistic has a t distribution with exactly $n_1 + n_2 - 2$ degrees of freedom. The pooled procedure can be seriously in error if the variances are not equal. The **POOLED** subcommand should not be used unless the variances are known to be equal.

The *F* Test for Equality of Variance

We will illustrate an *F* test on data of Example 7.7 of BPS. We will test

$$H_0: \sigma_1 = \sigma_2$$
$$H_a: \sigma_1 \neq \sigma_2$$

The hypothesis of equal spread can be tested in Minitab using an *F* test. The *F* test is not recommended for distributions that are not normal. Before we calculate the *F* statistic, it is important to verify that the distributions are normal. This is done graphically with a **STEMPLOT** or **HISTOGRAM**. **BOXPLOTS** are also useful for visually checking whether the variances appear to be different. The boxplots (on page 127) for the 2 week and 16 week data show that the distributions do not appear to have the same variances.

The *F* statistic is the ratio of the sample variances,

$$F = \frac{s_1^2}{s_2^2}$$

with the larger sample variance in the numerator. The test statistic can be computed with Minitab using Minitab's calculator or the **LET** command as illustrated below.

```
MTB > Let k1 = (STDEV(c2)/STDEV(c1))**2
MTB > print k1
```

Data Display

```
K1    12.2075
```

Once the *F* statistic has been calculated, the **CDF** command with the **F** subcommand can be used to find the *P*-value of the observations. The subcommand format is

```
F df numerator = K, df denominator = K
```

Below, we use the **CDF** command with the **F** subcommand to find $P(F < 12.2075)$.

```
MTB > cdf k1;
SUBC> f 4 4.
```

Cumulative Distribution Function

```
F distribution with 4 DF in numerator and 4 DF in denominator

      x        P( X <= x)
 12.2075         0.9837
```

Since the test is two-sided, the P-value is equal to $2 \times P(F < 12.2075)$. In the example above, the P-value is $2 \times (1 - 0.9837) = 0.0326$. Therefore, the difference between the spread on the two tests is statistically significant.

The **INVCDF** command can be used with the **F** subcommand to find the critical value of F that would be required to reject the null hypothesis at the 5% significance level. Since the test is two-sided, the probability in each tail must be 0.025 or less, so we find the value of F that has probability 0.975 of being less.

```
MTB > invcdf .975;
SUBC> f 4 4.
```

Inverse Cumulative Distribution Function

```
F distribution with 4 DF in numerator and 4 DF in denominator

P( X <= x)           x
  0.9750          9.6045
```

Therefore, for the sample sizes shown above, the F statistic would have to be at least 9.6 to reject H_0 at the 5% level.

EXERCISES

7.9   The design of controls and instruments has a large effect on how easily people can use them. A student project investigated this effect by asking 25 right-handed students to turn a knob (with their right hands) that moved an indicator by screw action. There were two identical instruments, one with a right-hand thread (the knob turns clockwise) and the other with a left-hand thread (the knob must be turned counterclockwise). The table below and TA07-02.MTW give the times required (in seconds) to move the indicator a fixed distance.

Subject	Right thread	Left thread	Subject	Right thread	Left thread
1	113	137	14	107	87
2	105	105	15	118	166
3	130	133	16	103	146
4	101	108	17	111	123
5	138	115	18	104	135
6	118	170	19	111	112
7	87	103	20	89	93
8	116	145	21	78	76
9	75	78	22	100	116
10	96	107	23	89	78
11	122	84	24	85	101
12	103	148	25	88	123
13	116	147			

(a) Each of the 25 students used both instruments. Discuss briefly how you would use randomization in arranging the experiment.

(b) The project hoped to show that right-handed people find right-hand threads easier to use. What is the parameter μ for a matched pairs t test? State H_0 and H_a in terms of μ.

(c) Use the **PAIR** command to carry out a test of your hypotheses. Give the *P*-value and report your conclusions.

7.10 Give a 90% **TINTERVAL** for the mean time advantage of right-hand over left-hand threads in the setting of Exercise 7.9. Do you think that the time saved would be of practical importance if the task were performed many times—for example, by an assembly-line worker? To help answer this question, find the mean time for right-hand threads as a percent of the mean time for left-hand threads.

7.13 Stored in EX07-13.MTW and given below are measurements (in millimeters) of a critical dimension for 16 auto engine crankshafts.

224.120	224.001	224.017	223.982	223.989	223.961
223.960	224.089	223.987	223.976	223.902	223.980
224.098	224.057	223.913	223.999		

The mean dimension is supposed to be 224 mm and the variability of the manufacturing process is unknown. Is there evidence that the mean dimension is not 224 mm?

(a) Check the data graphically for outliers or strong skewness that might threaten the validity of the *t* procedures. What do you conclude?

(b) Do these data give evidence that the process mean is not equal to the target value 224 mm? Use the **TTEST** command to find the exact *P*-value for the following hypothesis test.

$$H_0: \mu = 224$$
$$H_a: \mu \neq 224$$

7.16 Great white sharks are big and hungry. Here and in EX07-44.MTW are the lengths in feet of 44 great whites. Use 1-Sample *t* confidence interval and test to examine the data.

18.7	12.3	18.6	16.4	15.7	18.3	14.6	15.8	14.9	17.6	12.1
16.4	16.7	17.8	16.2	12.6	17.8	13.8	12.2	15.2	14.7	12.4
13.2	15.8	14.3	16.6	9.4	18.2	13.2	13.6	15.3	16.1	13.5
19.1	16.2	22.8	16.8	13.6	13.2	15.7	19.7	18.7	13.2	16.8

(a) Select the Graph button to examine these data for shape, center, spread, and outliers. The distribution is reasonably normal except for one outlier in each direction. Because these are not extreme and preserve the symmetry of the distribution, use of the *t* procedures is safe with 44 observations.

(b) Give a 95% confidence interval for the mean length of great white sharks. Based on this interval, is there significant evidence at the 5% level to reject the claim "Great white sharks average 20 feet in length"?

(c) It isn't clear exactly what parameter μ you estimated in (b). What information do you need to say what μ is?

7.21 Differences of electric potential occur naturally from point to point on a body's skin. Is the natural electric field strength best for helping wounds to heal? If so, changing the field will slow healing. The research subjects are anesthetized newts. Make a razor cut in both hind limbs. Let one heal naturally (the control). Use an electrode to change the electric field in the other to half its normal value. After two hours, measure the healing rate. TA07-04.MTW and Table 7.4 in BPS give the healing rates (in micrometers per hour) for 14 newts.

(a) As is usual, the paper did not report these raw data. Readers are expected to be able to interpret the summaries that the paper did report. The paper summarized the differences in the table above as

"−5.71 ± 2.82" and said, "All values are expressed as means ± standard error of the mean." Use the **PAIR** command to show where the numbers −5.71 and 2.82 come from.

(b) The researchers want to know if changing the electric field reduces the mean healing rate for all newts. State hypotheses, and based on the output in (a), give your conclusion.

(c) Give a 90% confidence interval for the amount by which changing the field changes the rate of healing. Then explain in a sentence what it means to say that you are "90% confident" of your result.

7.24 Makers of generic drugs must show that they do not differ significantly from the "reference" drug that they imitate. One aspect in which drugs might differ is their extent of absorption in the blood. TA07-05.MTW and Table 7.5 in BPS give data taken from 20 healthy nonsmoking male subjects for one pair of drugs. This is a matched pairs design. Subjects 1 to 10 received the generic drug first, and Subjects 11 to 20 received the reference drug first. In all cases, a washout period separated the two drugs so that the first had disappeared from the blood before the subject took the second. The subject numbers in the table were assigned at random to decide the order of the drugs for each subject answer the following questions.

(a) Do a data analysis of the differences between the absorption measures for the generic and reference drugs. Is there any reason not to apply *t* procedures?

(b) Use the **PAIR** command to a *t* test to answer the key question: Do the drugs differ significantly in absorption?

7.35 Ordinary corn doesn't have as much of the amino acid lysine as animals need in their feed. Plant scientists have developed varieties of corn that have increased amounts of lysine. In a test of the quality of high-lysine corn as animal feed, an experimental group of 20 one-day-old male chicks ate a ration containing the new corn. A control group of another 20 chicks received a ration that was identical except that it contained normal corn. Below and in EX07-35.MTW are the weight gains (in grams) after 21 days.

	Control			Experimental			
380	321	366	356	361	447	401	375
283	349	402	462	434	403	393	426
356	410	329	399	406	318	467	407
350	384	316	272	427	420	477	392
345	455	360	431	430	339	410	326

(a) Present the data graphically. Are there outliers or strong skewness that might prevent the use of t procedures?

(b) State the hypotheses for a statistical test of the claim that chicks fed high-lysine corn gain weight faster. Carry out the test using the **TWOT** command. Is the result significant at the 10% level? At the 5% level? At the 1% level?

(c) Give a 95% confidence interval for the mean extra weight gain in chicks fed high-lysine corn.

7.36 The Survey of Study Habits and Attitudes (SSHA) is a psychological test that measures the motivation, attitude toward school, and study habits of students. Scores range from 0 to 200. A selective private college gives the Survey of Study Habits and Attitudes (SSHA) to an SRS of both male and female freshmen. The data for both men and women are given in EX07-36.MTW. The data for the women are as follows:

| 154 | 109 | 137 | 115 | 152 | 140 | 154 | 178 | 101 |
| 103 | 126 | 126 | 137 | 165 | 165 | 129 | 200 | 148 |

Here are the scores of the men:

| 108 | 140 | 114 | 91 | 180 | 115 | 126 | 92 | 169 | 146 |
| 109 | 132 | 75 | 88 | 113 | 151 | 70 | 115 | 187 | 104 |

(a) Examine each sample graphically, with special attention to outliers and skewness. Is use of a t procedure acceptable for these data?

(b) Most studies have found that the mean SSHA score for men is lower than the mean score in a comparable group of women. Use the **TWOT** command to test this supposition here. Obtain a P-value, and give your conclusions.

(c) Give a 90% confidence interval for the mean difference between the SSHA scores of male and female freshmen at this college. This can also be done with the **TWOT** command by specifying a 90% level.

7.43 A study of computer-assisted learning examined the learning of "Blissymbols" by children. Blissymbols are pictographs (think of Egyptian hieroglyphs) that are sometimes used to help learning-impaired children communicate. The researcher designed two computer lessons that taught the same content using the same examples. One lesson required the children to interact with the material, while in the other the children controlled only the pace of the lesson. Call these two styles "Active" and "Passive." After the lesson, the computer presented a quiz that asked the children to identify 56 Blissymbols. Here and in EX07-43.MTW are the numbers of correct identifications by the 24 children in the Active and Passive groups:

Active	29	28	24	31	15	24	27	23	20	22	23	21
	24	35	21	24	44	28	17	21	21	20	28	16
Passive	16	14	17	15	26	17	12	25	21	20	18	21
	20	16	18	15	26	15	13	17	21	19	15	12

(a) Is there good evidence that active learning is superior to passive learning? State hypotheses, use the **TWOT** command to give the P-value. State your conclusion.

(b) Give a 90% confidence interval for the mean number of Blissymbols identified correctly in a large population of children after the Active computer lesson.

(c) What assumptions do your procedures in (a) and (b) require? Which of these assumptions can you use the data to check? Examine the data to check the assumptions and report your results.

7.44 Here and in EX07-44.MTW are the IQ test scores of 31 seventh-grade girls and 47 boys in a Midwest school district:

Girls

114	100	104	89	102	91	114	114	103	105	108
130	120	132	111	128	118	119	86	72	111	
103	74	112	107	103	98	96	112	112	93	

Boys

111	107	100	107	115	111	97	112	104	106	113	109
113	128	128	118	113	124	127	136	106	123	124	126
116	127	119	97	102	110	120	103	115	93	123	79
119	110	110	107	105	105	110	77	90	114	106	

(a) Find the mean scores for girls and for boys. It is common for boys to have somewhat higher scores on standardized tests. Is that true here?

136   Chapter 7

(b) Make **STEMPLOTS** or **HISTOGRAMS** of both sets of data. Because the distributions are reasonably symmetric with no extreme outliers, the *t* procedures will work well.

(c) Treat these data as SRSs from all seventh-grade students in the district. Is there good evidence that girls and boys differ in their mean IQ scores?

(d) Give a 90% confidence interval for the difference between the mean IQ scores of all boys and girls in the district.

(e) What other information would you ask for before accepting the results as describing all seventh-graders in the school district?

7.58 A selective private college gives the Survey of Study Habits and Attitudes (SSHA) to an SRS of both male and female freshmen. SSHA scores are generally less variable among women than among men. We want to know whether this is true for this college. The data are given in Exercise 7.36 and EX07-36.MTW.

(a) Verify graphically that the SSHA distributions are close to normally distributed.

(b) Use the **LET** command to compute the test statistic for an *F* test. (The numerator s^2 belongs to the group that H_a claims to have the larger σ.)

(c) Use the **CDF** command (no doubling of *p*) to obtain the *P*-value. Be sure the degrees of freedom are in the proper order. What do you conclude about the variation in SSHA scores?

7.64 Nitrites are often added to meat products as preservatives. In a study of the effect of nitrites on bacteria, researchers measured the rate of uptake of an amino acid for 60 cultures of bacteria: 30 growing in a medium to which nitrites had been added and another 30 growing in a standard medium as a control group. Tables 7.7 and TA07-07.MTW give the data from this study. Examine each of the two samples and briefly describe their distribution. Use the **TWOT** command to carry out a test of the research hypothesis that nitrites decrease amino acid uptake, and report your results.

7.66 A pharmaceutical manufacturer does a chemical analysis to check the potency of products. The standard release potency for cephalothin crystals is 910. An assay of 16 lots gives the following potency data. The data are also given in EX07-66.

897	914	913	906	916	918	905	921
918	906	895	893	908	906	907	901

- (a) Check the data for outliers or strong skewness that might threaten the validity of the *t* procedures.
- (b) Give a 95% **TINTERVAL** for the mean potency.
- (c) Use the **TTEST** command to determine whether there is significant evidence at the 5% level that the mean potency is not equal to the standard release potency.

7.71 Exercise 1.41 in BPS and EX01-41.MTW give 29 measurements of the density of the earth, made in 1798 by Henry Cavendish. Display the data graphically to check for skewness and outliers. Use the **TINTERVAL** command to give an estimate for the density of the earth from Cavendish's data and a margin of error for your estimate.

7.75 We want to compare the level of particulate in the city with the rural level on the same day. We suspect that pollution is higher in the city, and we hope that a statistical test will show that there is significant evidence to confirm this suspicion. TA07-08.MTW gives data on the concentration of airborne particulate matter in a rural area upwind from a small city and in the center of the city. Make a graph to check for conditions that might prevent the use of the test you plan to employ. Your graph should reflect the type of procedure you will use. Then carry out a significance test and report your conclusion. Also estimate the mean amount by which the city particulate level exceeds the rural level on the same day.

Chapter 8
Inference for Proportions

Commands to be covered in this chapter:

```
PONE  C...C or K K...K
PTWO  C...C or K K...K
```

The PONE Command

The **PONE** command is used to compute a confidence interval and hypothesis test of the proportion. The data format can be either raw or summarized. The Command format is

```
PONE   C...C or K K...K
```

In example 8.6 of BPS, Buffon tossed a coin 4040 times. He got 2048 heads. We can use the **PONE** command either by specifying a column of data (C1) with 2048 heads and 1992 tails as shown below.

```
MTB > pone c1
```

Test and Confidence Interval for One Proportion

```
Test of p = 0.5 vs p not = 0.5

Success = tail

                                                         Exact
Variable       X       N   Sample p        95.0 % CI    P-Value
C1          1992    4040   0.493069   (0.477539, 0.508610)  0.387
```

Alternatively, the data can be summarized with the command:

```
MTB > pone 4040 2048
```

Test and Confidence Interval for One Proportion

Test of p = 0.5 vs p not = 0.5

Sample	X	N	Sample p	95.0 % CI	Exact P-Value
1	1992	4040	0.493069	(0.477539, 0.508610)	0.387

Note that the **PONE** command computes confidence interval and performs a hypothesis test of the proportion By default, Minitab uses an exact method to calculate the test probability. If you choose to use a normal approximation, Minitab calculates the confidence interval as

$$\hat{p} \pm z * \sqrt{\frac{\hat{p}(1-\hat{p})}{n}}$$

and calculates the test statistic (z) as:

$$z = \frac{\hat{p} - p_0}{\sqrt{\frac{p_0(1-p_0)}{n}}}$$

where $\hat{p}$ is the observed probability equal to x/n, where x is the observed number of successes in n trials, and p_0 is the hypothesized probability. The probabilities are obtained from a standard normal distribution table. When p_0 is not specified in Test proportion, $p_0 = .5$ is used. Minitab performs a two-tailed test unless you specify a one-tailed test. The default value for the confidence interval is 95%. Other values can be specified using **CONFIDENCE**, **ALTERNATIVE**, **TEST**, and **USEZ** subcommands. The formats are:

```
CONFIDENCE level is K
ALTERNATIVE = K
TEST the null hypothesis
USEZ
```

Selecting **Stat ➤ Basic Statistics ➤ 1 Proportion** is another way to compute a confidence interval or perform a hypothesis test for a proportion. The dialog box allows for either of the data formats described above.

Inference for Proportions 141

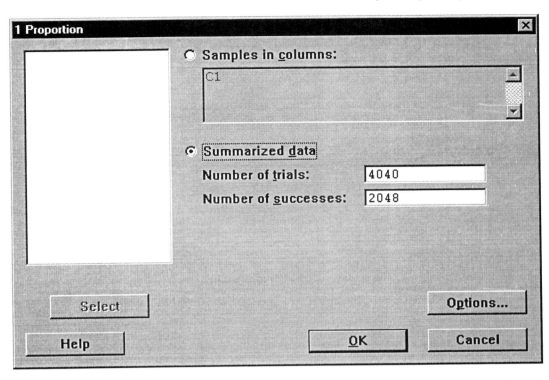

The options subdialog box allows the user to specify the confidence level, test proportion, alternative hypothesis, and to check a box to specifying that Minitab use tests and intervals based on normal distributions.

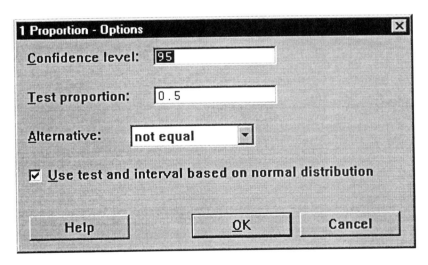

Below we specify that the test and confidence interval be based on the normal distribution. In this case, the results are slightly different. The difference is more significant when n is small.

```
MTB > POne 4040 2048;
SUBC>    Confidence 95;
SUBC>    Test 0.5;
SUBC>    Alternative 0;
SUBC>    UseZ.
```

Test and Confidence Interval for One Proportion

Test of p = 0.5 vs p not = 0.5

```
Sample    X      N   Sample p        95.0 % CI         Z-Value   P-Value
1       2048   4040   0.506931   (0.491514, 0.522347)    0.88     0.378
```

The PTWO Command

The **PTWO** command performs a test of two binomial proportions. We use **PTWO** to compute a confidence interval and perform a hypothesis test of the difference between two proportions. Suppose you wanted to know whether a drug (gemfibrozil) will reduce heart attacks. The Helsinki Heart Study described in Example 8.11 of BPS looked at this question. 2051 men took Gemfibrozil to reduce their cholesterol levels and 2030 men took a placebo. 56 men in the gemfibrozil group and 84 men in the placebo group had heart attacks. We will use the **PTWO** command to test

$$H_0 : p_1 = p_2$$
$$H_a : p_1 < p_2.$$

The **PTWO** command calculates the confidence interval as

$$\hat{p}_1 - \hat{p}_2 \pm z^* \sqrt{\frac{\hat{p}_1(1-\hat{p}_1)}{n_1} + \frac{\hat{p}_2(1-\hat{p}_2)}{n_2}},$$

where $\hat{p}_1$ and $\hat{p}_2$ are the observed probabilities of sample one and sample two respectively, $\hat{p} = x/n$, where x is the observed success in n trials. You can use the **CONFIDENCE** subcommand to specify a confidence level. The confidence level is 95% by default. The calculation of the test statistic, z, depends on the method used to estimate p. By default, Minitab uses separate estimates of p for each population and calculates z by:

$$z = \frac{\hat{p}_1 - \hat{p}_2}{\sqrt{\frac{\hat{p}_1(1-\hat{p}_1)}{n_1} + \frac{\hat{p}_2(1-\hat{p}_2)}{n_2}}}$$

This test statistic is different from the one described in Chapter 8 of BPS. To calculate z as described in the BPS, always use the **POOLED** subcommand to use a pooled estimate of p for the test. In this case, Minitab calculates z by:

$$z = \frac{\hat{p}_1 - \hat{p}_2}{\sqrt{\hat{p}(1-\hat{p})\left(\frac{1}{n_1} + \frac{1}{n_2}\right)}}, \text{ where } \hat{p} = \frac{x_1 + x_2}{n_1 + n_2}.$$

Minitab performs a two-tailed test unless you use the **ALTERNATIVE** subcommand to specify a one-tailed test. In the example below, we use the **CONFIDENCE, ALTERNATIVE,** and **POOLED** subcommands. Note that on the command line we enter the summary data: $n_1, x_1, n_2,$ and x_2, in that order.

```
MTB > PTwo 2051 56 2030 84;
SUBC>    Confidence 95.0;
SUBC>    Alternative -1;
SUBC>    Pooled.
```

Test and Confidence Interval for Two Proportions

```
Sample      X       N      Sample p
1          56     2051    0.027304
2          84     2030    0.041379

Estimate for p(1) - p(2):  -0.0140756
95% CI for p(1) - p(2):   (-0.0252472, -0.00290387)
Test for p(1) - p(2) = 0 (vs < 0):  Z = -2.47   P-Value = 0.007
```

The output shows that the P value is 0.007. Because $P < 0.01$, the results are statistically significant at the $\alpha = 0.01$ level. There is strong evidence that gemfibrozil reduced the rate of heart attacks.

The same calculation can also be obtained by selecting **Stat ➤ Basic Statistics ➤ 2 Proportions** from the menu. The summary data can be entered in the dialog box by checking Summarized data.

Chapter 8

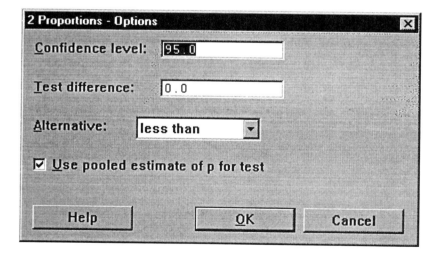

The options subdialog box allows the user to specify the confidence level, test proportion, alternative hypothesis, and to check a box to specifying that Minitab use a pooled estimated of p for the test.

EXERCISES

8.10 *The New York Times* and CBS News conducted a nationwide poll of 1048 randomly selected 13- to 17-year-olds. Of these teenagers, 692 had televisions in their rooms and 189 named Fox as their favorite television network. We will act as if the sample were an SRS.

(a) Use the **PONE** command with the **USEZ** subcommand to give 95% confidence intervals for the proportion of all people in this age group who have televisions in their rooms and the proportion who would choose Fox as their favorite network. Check that we can use our methods.

(b) The news article says, "In theory, in 19 cases out of 20, the poll results will differ by no more than three percentage points in either direction from what would have been obtained by seeking out all American teenagers." Explain how your results agree with this statement.

(c) Is there good evidence that more than half of all teenagers have televisions in their room? State hypotheses, give a test statistic, and state your conclusion about the strength of the evidence.

8.11 In a recent year, 73% of first-year college students responding to a national survey identified "being very well-off financially" as an important personal goal. A state university finds that 132 of an SRS of 200 of its first-year students say that this goal is important.

(a) Use the **PONE** command with the **USEZ** subcommand to give a 95% confidence interval for the proportion of all first-year students at the university who would identify being well-off as an important personal goal.

(b) Is there good evidence that the proportion of first-year students at this university who think being very well-off is important differs from the national value, 73%? (Be sure to state hypotheses, give the *P*-value, and state your conclusion.)

(c) Check that you can safely use the methods of this section in both (a) and (b).

8.21 One-sample procedures for proportions, like those for means, are used to analyze data from matched pairs designs. Here is an example. Each of 50 subjects tastes two unmarked cups of coffee and says which he or she prefers. One cup in each pair contains instant coffee; the other, fresh-brewed coffee. Thirty-one of the subjects prefer the fresh-brewed coffee. Take *p* to be the proportion of the population who would prefer fresh-brewed coffee in a blind tasting.

(a) Use the **PONE** command to test the claim that a majority of people prefer the taste of fresh-brewed coffee. State hypotheses and report the z statistic, and its P-value. Is your result significant at the 5% level? What is your practical conclusion?

(b) Find a 90% confidence interval for p.

(c) When you do an experiment like this, in what order should you present the two cups of coffee to the subjects?

8.24 A study of injuries to in-line skaters used data from the National Electronic Injury Surveillance System, which collects data from a random sample of hospital emergency rooms. In the six-month study period, 206 people came to the sample hospitals with injuries from in-line skating. We can think of these people as an SRS of all people injured while skating. Researchers were able to interview 161 of these people. Wrist injuries (mostly fractures) were the most common.

(a) The interviews found that 53 people were wearing wrist guards and 6 of these had wrist injuries. Of the 108 who did not wear wrist guards, 45 had wrist injuries. What are the two sample proportions of wrist injuries?

(b) Use the **PTWO** command to give a 95% confidence interval for the difference between the two population proportions of wrist injuries. State carefully what populations your inference compares. We would like to draw conclusions about all in-line skaters, but we have data only for injured skaters.

(c) What was the percent of nonresponse among the original sample of 206 injured skaters? Explain why nonresponse may bias your conclusions.

8.26 The 1958 Detroit Area Study was an important investigation of the influence of religion on everyday life. The sample "was basically a simple random sample of the population of the metropolitan area" of Detroit, Michigan. Of the 656 respondents, 267 were white Protestants and 230 were white Catholics. The study took place at the height of the Cold War. One question asked if the right of free speech included the right to make speeches in favor of communism. Of the 267 white Protestants, 104 said "Yes," while 75 of the 230 white Catholics said "Yes," Use the **PTWO** command to give a 95% confidence interval for the difference between the proportion of Protestants who agreed that communist speeches are protected and the proportion of Catholics who held this opinion.

8.30 Telephone surveys often have high rates of nonresponse. When the call is handled by an answering machine, perhaps leaving a message on the machine will encourage people to respond when they are called again. Here are data from a study in which (at random) a message was or was not left when an answering machine picked up the first call from a survey:

	Total households	Eventual contact	Completed survey
No message	100	58	33
Message	291	200	134

(a) Use the **PTWO** command and the **POOLED** subcommand to see whether there is good evidence that leaving a message increases the proportion of households that are eventually contacted.

(b) Is there good evidence that leaving a message increases the proportion who complete the survey?

(c) If you find significant effects, look at their size. Do you think these effects are large enough to be important to survey takers?

8.31 Exercise 8.18 in BPS describes a study of whether patients who file complaints leave a health maintenance organization (HMO). We want to know whether complainers are more likely to leave than patients who do not file complaints. In the year of the study, 639 patients filed complaints, and 54 of these patients left the HMO voluntarily. For comparison, the HMO chose an SRS of 743 patients who had not filed complaints. Twenty-two of these patients left voluntarily. How much higher is the proportion of complainers who leave? Use the **PTWO** command to give a 90% confidence interval.

8.32 The drug AZT was the first drug that seemed effective in delaying the onset of AIDS. Evidence for AZT's effectiveness came from a large randomized comparative experiment. The subjects were volunteers who were infected with HIV, the virus that causes AIDS, but did not yet have AIDS. The study assigned 435 of the subjects at random to take 500 milligrams of AZT each day, and another 435 to take a placebo. At the end of the study, 38 of the placebo subjects and 17 of the AZT subjects had devel-

148   *Chapter 8*

oped AIDS. We want to test the claim that taking AZT lowers the proportion of infected people who will develop AIDS in a given period of time.

(a) State hypotheses, and check that you can safely use the z procedures.

(b) Use the **PTWO** command and the **POOLED** subcommand to determine how significant is the evidence that AZT is effective.

(c) The experiment was double-blind. Explain what this means.

8.36 Different kinds of companies compensate their key employees in different ways. Established companies may pay higher salaries, while new companies may offer stock options that will be valuable if the company succeeds. Do high-tech companies tend to offer stock options more often than other companies? One study looked at a random sample of 200 companies. Of these, 91 were listed in the *Directory of Public High Technology Corporations* and 109 were not listed. Treat these two groups as SRSs of high-tech and non-high-tech companies. Seventy-three of the high-tech companies and 75 of the non-high-tech companies offered incentive stock options to key employees.

(a) Is there evidence that a higher proportion of high-tech companies offer stock options? Use the **PTWO** command and the **POOLED** subcommand to determine whether the evidence is significant.

(b) Give a 95% confidence interval for the difference in the proportions of the two types of companies that offer stock options.

8.41 Never forget that even small effects can be statistically significant if the samples are large. To illustrate this fact, consider a study of small-business failures. The study looked at 148 food-and-drink businesses in central Indiana.

(a) Of these, men headed 106 and women headed 42. During a three-year period, 15 of the men's businesses and 7 of the women's businesses failed. These sample proportions are quite close to each other. Use the **PTWO** command and the **POOLED** subcommand to see whether there is a significant difference. Give the *P*-value for the z test of the hypothesis that the same proportion of women's and men's businesses fail. (Use the two-sided alternative.)

(b) Now suppose that the same sample proportions came from a sample 30 times as large. That is, 210 out of 1260 businesses headed by women and 450 out of 3180 businesses headed by men fail. Verify that the proportions of failures are exactly the same as in (a). Repeat the z test for the new data, and show that it is now significant at the 5% level.

(c) It is wise to use a confidence interval to estimate the size of an effect, rather than just giving a P-value. Give 95% confidence intervals for the difference between the proportions of women's and men's businesses that fail for the settings of both (a) and (b). What is the effect of larger samples on the confidence interval?

8.47 Some people think that chemists are more likely than other parents to have female children. (Perhaps chemists are exposed to something in their laboratories that affects the sex of their children.) The Washington State Department of Health lists the parents' occupations on birth certificates. Between 1980 and 1990, 555 children were born to fathers who were chemists. Of these births, 273 were girls. During this period, 48.8% of all births in Washington State were girls. Is there evidence that the proportion of girls born to chemists is higher than the state proportion? Use the **PONE** command specifying that the test proportion is .488.

8.49 A study of "adverse symptoms" in users of over-the-counter pain relief medications assigned subjects at random to one of two common pain relievers: acetaminophen and ibuprofen. (Both of these pain relievers are sold under various brand names, sometimes combined with other ingredients.) In all, 650 subjects took acetaminophen, and 44 experienced some adverse symptom. Of the 347 subjects who took ibuprofen, 49 had an adverse symptom. How strong is the evidence that the two pain relievers differ in the proportion of people who experience an adverse symptom?

(a) State hypotheses and check that you can use the z test.

(b) Use the **PTWO** command and the **POOLED** subcommand to find the P-value of the test and give your conclusion.

Chapter 9
Inference for Two-way Tables

Command to be covered in this chapter:

`CHISQUARE test on table stored in C...C`

The CHISQUARE Command

The `CHISQUARE` command does a χ^2 test of the null hypothesis that there is "no relationship" between the column variable and the row variable in a two-way table. The command format is

`CHISQUARE test on table stored in C...C`

The command performs a χ^2 test for independence or association on a frequency table that has already been formed and stored in the worksheet. You may specify up to seven columns. The columns must contain integer values. Below, we will illustrate the `CHISQUARE` command on the data from Example 9.1 in BPS. The 72 subjects are classified according to their treatment (Desipramine, Lithium, and Placebo) and whether or not they avoided relapse into cocaine use during a three-year study.

	Relapse		
Treatment	No	Yes	Total
Desipramine	14	10	24
Lithium	6	18	24
Placebo	4	20	24
Total	24	48	72

We want to test the null hypothesis that there are no differences among the proportions of successes for addicts given the three treatments. The alternative is that there is some difference, that not all three proportions are equal.

$$H_0: p_1 = p_2 = p_3$$
$$H_a: \text{not all of } p_1, p_2, \text{ and } p_3 \text{ are equal}$$

The data are entered below and then the **CHISQUARE** command is illustrated.

```
MTB > set c1
DATA> 14 6 4
DATA> set c2
DATA> 10 18 20
DATA> end
MTB > name c1 'no' c2 'yes'
MTB > chis c1 c2
```

Chi-Square Test

Expected counts are printed below observed counts

```
            no      yes     Total
   1        14       10       24
           8.00    16.00

   2         6       18       24
           8.00    16.00

   3         4       20       24
           8.00    16.00

Total       24       48       72

Chi-Sq =   4.500 +  2.250 +
           0.500 +  0.250 +
           2.000 +  1.000  = 10.500
DF = 2, P-Value = 0.005
```

The **CHISQUARE** command provides the χ^2 statistic, the number of degrees of freedom and the *P*-value. The number of degrees of freedom for the χ^2 statistic is equal to $(\text{rows}-1) \times (\text{columns}-1)$. In this example, the *P*-value is equal to 0.005. The small *P*-value gives us good reason to conclude that there *are* differences between the three treatments.

The command can also be used by selecting **Stat ➤ Tables ➤ Chi-Square Test** from the menu. Only the columns containing the table need to be entered in the dialog box before clicking OK.

Another use for the chi-square test is to test whether data show a statistically significant relationship between two categorical variable. Example 9.7 in BPS describes a study of the relationship between men's marital status and the levels of their jobs. Each man's job has a grade set the company that reflects the value to the company. The data are shown in the Data window below: We will use **CHISQUARE** to test the null hypothesis

H_0: there is no relationship between marital status and job grade.

	C1	C2	C3	C4	C5
	single	married	divorced	widowed	
1	58	874	15	8	
2	222	3927	70	20	
3	50	2396	34	10	
4	7	533	7	4	
5					
6					

```
MTB > ChiSquare  'single' 'married' 'divorced' 'widowed'.
```

Chi-Square Test

```
Expected counts are printed below observed counts

        single  married  divorced  widowed   Total
   1        58      874        15        8     955
         39.08   896.44     14.61     4.87

   2       222     3927        70       20    4239
        173.47  3979.05     64.86    21.62

   3        50     2396        34       10    2490
        101.90  2337.30     38.10    12.70

   4         7      533         7        4     551
         22.55   517.21      8.43     2.81

Total      337     7730       126       42    8235

Chi-Sq =   9.158 +   0.562 +   0.010 +   2.011 +
          13.575 +   0.681 +   0.407 +   0.121 +
          26.432 +   1.474 +   0.441 +   0.574 +
          10.722 +   0.482 +   0.243 +   0.504 = 67.397
DF = 9, P-Value = 0.000
2 cells with expected counts less than 5.0
```

The observed $\chi^2 = 67.397$ is very large. The *P*-value, is approximately zero, so we have strong evidence that job grade is related to marital status.

154   Chapter 9

The Minitab output warns us that the expected counts in two of the 16 cells are less than five. You can safely use the chi-square test when no more than 20% of the expected counts are less than five and all individual counts are one or greater.

EXERCISES

9.2   How are the smoking habits of students related to their parents' smoking? Here are data from a survey of students in eight Arizona high schools.

	Student smokes	Student does not smoke
Both parents smoke	400	1380
One parent smokes	416	1823
Neither parent smokes	188	1168

(a)   Explain in words what the null hypothesis $H_0: p_1 = p_2 = p_3$ says about student smoking.

(b)   Enter the data and use the **CHISQUARE** command to find the expected counts if H_0 is true, and display them in a two-way table with the observed counts.

(c)   Compare the table of observed and expected counts. Explain how the comparison expresses the association between parent smoking and student smoking.

9.4   In Exercise 9.2, you began to analyze data on the relationship between smoking by parents and smoking by high school students.

(a)   Using the Minitab output from Exercise 9.2, what is the value of the χ^2 statistic? What are its degrees of freedom?

(b)   What is the P-value for the test? Explain in simple language what it means to reject H_0 in this setting.

(c)   Does this study convince you that parent smoking *causes* student smoking? Explain your answer.

9.8   Exercise 8.18 in BPS compared HMO members who filed complaints with an SRS of members who did not complain. The study actually broke the complainers into two subgroups: those who filed complaints about medical treatment and those who filed nonmedical complaints. Here are the data on the total number in each group and the number who voluntarily left the HMO:

Inference for Two-way Tables

	No complaint	Medical complaint	Nonmedical complaint
Total	743	199	440
Left	22	26	28

(a) Enter the data into a Minitab worksheet. Use the **LET** command to find the number of child-care workers in each group that stayed.

(b) Can we safely use the chi-square test? What null and alternative hypotheses does χ^2 test?

(c) Use the **CHISQUARE** command to test do the χ^2 test.

(d) What is the χ^2 statistic? What are its degrees of freedom? What is the P-value?

(e) What do you conclude from these data?

9.9 Gastric freezing was once a recommended treatment for ulcers in the upper intestine. Use of gastric freezing stopped after experiments showed it had no effect. One randomized comparative experiment found that 28 of the 82 gastric freezing patients improved, while 30 of the 78 patients in the placebo group improved. We can test the hypothesis of "no difference" between the two groups in two ways: using the two-sample z statistic or using the chi-square statistic.

(a) State the null hypothesis with a two-sided alternative and use the **PTWO** command to carry out the z test. What is the P-value?

(b) Present the data in a 2 × 2 table. Use the **CHISQUARE** test to test the hypothesis from (a). Verify that the χ^2 statistic is the square of the z statistic and that the P-values agree.

(c) What do you conclude about the effectiveness of gastric freezing as a treatment for ulcers?

9.13 A study of the career plans of young women and men sent questionnaires to all 722 members of the senior class in the College of Business Administration at the University of Illinois. One question asked which major within the business program the student had chosen. Here are the data from the students who responded:

	Female	Male
Accounting	68	56
Administration	91	40
Economics	75	76
Finance	61	59

a) Use the **CHISQUARE** command to test the null hypothesis that there is no relation between the gender of students and their choice of major. Give a *P*-value and state your conclusion.

b) Describe the differences between the distributions of majors for women and men with percents, with a graph, and in words.

c) Which two cells have the largest components of the chi-square statistic? How do the observed and expected counts differ in these cells? (This should strengthen your conclusions in (b).)

d) Two of the observed cell counts are small. Do these data satisfy your guidelines for safe use of the chi-square test?

e) What percent of the students did not respond to the questionnaire? The nonresponse weakens conclusions drawn from these data.

9.15 Shopping at secondhand stores is becoming more popular and has even attracted the attention of business schools. A study of customers' attitudes toward secondhand stores interviewed samples of shoppers at two secondhand stores of the same chain in two cities. The breakdown of the respondents by sex is as follows.

	City 1	City 2
Men	38	68
Women	203	150
Total	241	218

Is there a significant difference between the proportions of women customers in the two cities?

(a) State the null hypothesis, find the sample proportions of women in both cities, do a two-sided *z* test, and give a *P*-value using the methods of Chapter 8.

(b) Use the **CHISQUARE** command to calculate the χ^2 statistic and show that it is the square of the *z* statistic. Show that the *P*-value agrees with your result from (a).

9.17 A large study of child care used samples from the data tapes of the Current Population Survey over a period of several years. The result is close to an SRS of child-care workers. The Current Population Survey has three classes of child-care workers: private household, nonhousehold, and preschool teacher. Here are data on the number of blacks among women workers in these three classes.

Inference for Two-way Tables

	Total	Black
Household	2455	172
Nonhousehold	1191	167
Teachers	659	86

(a) Enter the data into a Minitab worksheet. Use the **LET** command to find the number of child-care workers in each class who are not black.

(b) Can we safely use the chi-square test? What null and alternative hypotheses does χ^2 test?

(c) Use the **CHISQUARE** command to test do the χ^2 test.

(d) What is the χ^2 statistic? What are its degrees of freedom? What is the P-value?

(e) What do you conclude from these data?

9.23 In 1912 the luxury liner *Titanic*, on its first voyage across the Atlantic, struck an iceberg and sank. Some passengers got off the ship in lifeboats, but many died. Think of the *Titanic* disaster as an experiment in how the people of that time behaved when faced with death in a situation where only some can escape. The passengers are a sample from the population of their peers. Here is information about who lived and who died, by gender and economic status. (The data leave out a few passengers whose economic status is unknown.)

	Men			Women		
Status	Died	Survived	Status	Died	Survived
Highest	111	61	Highest	6	126
Middle	150	22	Middle	13	90
Lowest	419	85	Lowest	107	101
Total	680	168	Total	126	317

(a) Compare the percents of men and of women who died. Is there strong evidence that a higher proportion of men die in such situations? Why do you think this happened?

(b) Use the **CHISQUARE** command to see whether there is a relationship between survival and economic status. Describe how the three economic classes differ in the percent of women who died. Are these differences statistically significant?

(c) Now look only at the men and answer the same questions.

9.24 Example 2.23 of BPS presents artificial data that illustrate Simpson's paradox. The data given below concern patient outcomes in hospitals. Patients are classified by condition before the operation, hospital, and whether the patient survived at least 6 weeks following surgery. Here is a three-way table of the data.

	Good Condition			Poor Condition	
	Hospital A	Hospital B		Hospital A	Hospital B
Died	6	8	Died	57	8
Survived	594	592	Survived	1443	192

(a) Apply the **CHISQUARE** test to the data for all patients combined and summarize the results.

(b) Do separate **CHISQUARE** tests for the patients in good condition and for those in poor condition. Summarize these results.

(c) Are the effects that illustrate Simpson's paradox in this example statistically significant?

Chapter 10
One-way Analysis of Variance

Commands to be covered in this chapter:

```
ONEWAY data in C, levels in C
AOVONEWAY for samples in C...C
```

The ONEWAY Command

The **ONEWAY** command performs a one-way analysis of variance, with the response in one column and the subscripts in another column. The factor column may be numeric or text, and may contain any value. The levels do not need to be in any special order. The command format is given below.

```
ONEWAY data in C, levels in C
```

The **ONEWAY** command can also be used from the menu by selecting **Stat ➤ ANOVA ➤ One-way**.

The command is demonstrated on data from Example 10.1 of BPS. In Table 10.1 of BPS and in TA10-01.MTW, data are given on highway gas mileage (in miles per gallon as reported by the Environmental Protection Agency) for 28 midsize cars, 8 standard size pickup trucks, and 26 SUVs. The types (1 = midsize, 2 = truck, and 3 = SUV) are given in the first column. The gas mileage is given in the second column.

```
MTB > info
```

Information on the Worksheet

```
Column  Count  Name
C1         62  Type
C2         62  Mileage
```

160   *Chapter 10*

We want to compare the gas mileages for the three types of vehicles. Before proceeding with the **ONEWAY** command, it is important to check that the assumptions of one-way analysis of variance are satisfied. Specifically, the populations are normal with possibly different means and the same variance. **BOXPLOTS** are useful for visually checking these assumptions. We can produce side-by-side boxplots for this data, with the following command.

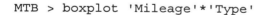

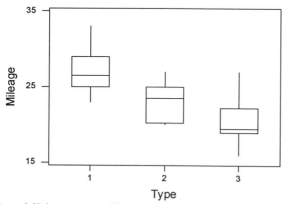

In addition, we will use the **DESCRIBE** command to summarize the data for the 3 types of vehicles.

```
MTB > desc c2;
SUBC> by c1.
```

Descriptive Statistics

Variable	Type	N	Mean	Median	TrMean	StDev
Mileage	1	28	27.107	26.500	27.038	2.629
	2	8	23.125	23.500	23.125	2.588
	3	26	20.423	19.500	20.333	2.914

Variable	Type	SE Mean	Minimum	Maximum	Q1	Q3
Mileage	1	0.497	23.000	33.000	25.000	29.000
	2	0.915	20.000	27.000	20.250	25.000
	3	0.572	16.000	27.000	19.000	22.250

Since the ratio of the largest to the smallest standard deviation is less than 2, it is safe to do one-way analysis of variance.

Call the mean highway gas mileages for the three types of vehicles μ_1 for midsize cars, μ_2 for pickups, and μ_3 for SUVs. We want to test the null hypothesis that there are *no differences* among the mean highway gas mileages for the three vehicle types:

$$H_0: \mu_1 = \mu_2 = \mu_3$$

The alternative is that there is some difference.

$$H_a: \text{not all of } \mu_1, \mu_2, \text{and } \mu_3 \text{ are equal}$$

The analysis of variance (ANOVA) output from the **ONEWAY** command is given below:

```
MTB > oneway c2 c1
```

One-way Analysis of Variance

```
Analysis of Variance for Mileage
Source     DF        SS        MS         F        P
Type        2    606.37    303.19     40.12    0.000
Error      59    445.90      7.56
Total      61   1052.27
                                     Individual 95% CIs For Mean
                                     Based on Pooled StDev
Level       N      Mean     StDev   ---+---------+---------+---------+---
1          28    27.107     2.629                              (---*----)
2           8    23.125     2.588             (-------*------)
3          26    20.423     2.914   (----*---)
                                    ---+---------+---------+---------+---
Pooled StDev =     2.749           20.0      22.5      25.0      27.5
```

The output provides the ANOVA table. The columns in this table are labeled Source, DF (degrees of freedom), SS (sum of squares), MS (mean square), F, and P. The rows in the table are labeled Type, Error, and Total. Consider our model

$$\text{DATA} = \text{FIT} + \text{RESIDUAL}$$

The Type row corresponds to the FIT term, the Error row corresponds to the RESIDUAL term, and the Total row corresponds to the DATA term. Notice that both the degrees of freedom and the sum of squares add to the value in the Total row.

The output provides the pooled standard deviation in the last line. It is given as equal to 2.749. Note that it can also be computed from the ANOVA table using the sum of squares and degrees of freedom for the Error row. That is,

$$s_p^2 = \frac{SS}{DF} = \frac{445.9}{59} = 7.558$$

which implies that $s_p = 2.749$.

The F statistic is given in the ANOVA table. If H_0 is true, the F statistic has an $F(\text{DFG}, \text{DFE})$ distribution, where DFG stands for degrees of freedom for groups and DFE stands for degrees of freedom for error. DFG = $I-1$, the number of groups minus 1. DFE = $N-I$, the number of observations minus the number of groups. The *P*-value for this distribution is also given above. In this example, the *P*-value is given as 0.000. This means that the *P*-value is 0 to three decimal places, or *P*-value < .001. This is strong evidence that the means are not all equal.

For information purposes, the output from the **ONEWAY** command provides the mean and standard deviation for each group and plots individual 95% confidence intervals for the means. Each confidence interval is of the form

$$\left(\bar{x}_i - t\frac{s_p}{\sqrt{n_i}}, \bar{x}_i + t\frac{s_p}{\sqrt{n_i}}\right)$$

where $\bar{x}_i$ and n_i are the sample mean and sample size for level *i*, s_p = Pooled StDev is the pooled estimate of the common standard deviation, and t^* is the value from a *t* table corresponding to 95% confidence and the degrees of freedom associated with MS Error. The truck and SUV intervals overlap slightly, and the midsize-car interval lies above them on the mileage scale. We conclude that the most important difference among the means is that midsize cars have better gas mileage than trucks and SUVs.

The AOVONEWAY Command

The **AOVONEWAY** command does exactly the same analysis as **ONEWAY**, but uses a different form of input. The data for each level must be in a separate column. There must be two or more levels (columns) specified. The same number of observations is not required in each column. The command format is given below.

```
AOVONEWAY for samples in C...C
```

The **AOVONEWAY** command can also be used from the menu by selecting **Stat ➤ ANOVA ➤ One-way (Unstacked)**. The output will be exactly the same as from the **ONEWAY** command except that the levels will be named instead of numbered.

```
MTB > AOVOneway 'Midsize' 'Truck' 'SUV'.
```

One-way Analysis of Variance

```
Analysis of Variance
Source      DF         SS         MS         F         P
Factor       2     606.37     303.19     40.12     0.000
Error       59     445.90       7.56
Total       61    1052.27
                                       Individual 95% CIs For Mean
                                       Based on Pooled StDev
Level        N       Mean      StDev   ---+---------+---------+---------+---
Midsize     28     27.107      2.629                                  (---*----)
Truck        8     23.125      2.588              (-------*------)
SUV         26     20.423      2.914   (----*---)
                                       ---+---------+---------+---------+---
Pooled StDev        2.749           20.0       22.5       25.0       27.5
```

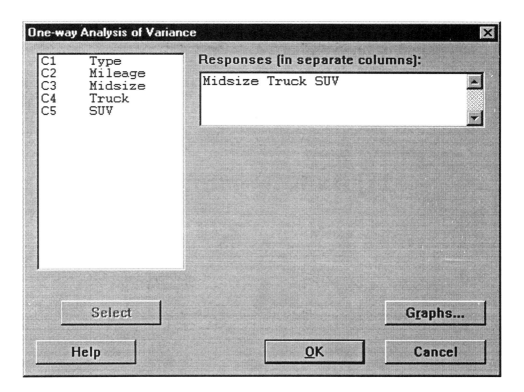

EXERCISES

10.1 If you are a dog lover, perhaps having your dog along reduces the effect of stress. To examine the effect of pets in stressful situations, researchers recruited 45 women who said they were dog lovers. Fifteen of the subjects were randomly assigned to each of three groups to do a stressful task

164   Chapter 10

alone, with a good friend present, or with their dog present. (The stressful task was to count backward by 13s or 17s.) The subject's mean heart rate during the task is one measure of the effect of stress. Table 10.2 in BPS and TA10-02.MTW contain the data.

(a) Make side-by-side **BOXPLOTS** of the heart rates for the three groups (round to the nearest whole number of beats). Do any of the groups show outliers or extreme skewness?

(b) Use the **ONEWAY** command for these data. Do the mean heart rates for the groups appear to show that the presence of a pet or a friend reduces heart rate during a stressful task?

(c) What are the values of the ANOVA F statistic and its P-value? What hypotheses does F test? Briefly describe the conclusions you draw from these data. Did you find anything surprising?

10.2 How much corn per acre should a farmer plant to obtain the highest yield? Too few plants will give a low yield. On the other hand, if there are too many plants, they will compete with each other for moisture and nutrients, and yields will fall. To find out, plant at different rates on several plots of ground and measure the harvest. (Be sure to treat all the plots the same except for the planting rate.) Here and in EX10-02.MTW are data from such an experiment:

Plants per acre	Yield (bushels per acre)			
12,000	150.1	113.0	118.4	142.6
16,000	166.0	120.7	135.2	149.8
20,000	165.3	130.1	139.6	149.9
24,000	134.7	138.4	156.1	
28,000	119.0	150.5		

(a) Make **STEM-AND-LEAF** displays of yield for each number of plants per acre. Use the **BY** subcommand so that all displays will be on the same scale. What do the data appear to show about the influence of plants per acre on yield?

(b) Use the **ONEWAY** to assess the statistical significance of the observed differences in yield. What are H_0 and H_a for the ANOVA F test in this situation?

(c) What is the sample mean yield for each planting rate? What does the ANOVA F test say about the significance of the effects you observe?

(d) The observed differences among the mean yields in the sample are quite large. Why are they not statistically significant?

10.16 Table 10.3 and TA10-03.MTW give data on the number of tree species per forest plot as well as on the number of individual trees. In the previous exercise, you examined the effect of logging on the number of trees. Use the **ONEWAY** command to analyze the effect of logging on the number of species.

(a) Use the **DESCRIBE** command to find the group means and standard deviations. Do the standard deviations satisfy our rule of thumb for safe use of ANOVA? What do the means suggest about the effect of logging on the number of species?

(b) Report the F statistic and its P-value and state your conclusion.

10.17 How do nematodes (microscopic worms) affect plant growth? A botanist prepares 16 identical planting pots and then introduces different numbers of nematodes into the pots. He transplants a tomato seedling into each plot. Here and in EX10-17.MTW are data on the increase in height of the seedlings (in centimeters) 16 days after planting.

Nematodes	Seedling growth			
0	10.8	9.1	13.5	9.2
1000	11.1	11.1	8.2	11.3
5000	5.4	4.6	7.4	5.0
10,000	5.8	5.3	3.2	7.5

(a) Use the **DESCRIBE** command to examine the means and standard deviations for the four treatments. What do the data appear to show about the effect of nematodes on growth?

(b) State H_0 and H_a for an ANOVA on these data.

(c) Use the **ONEWAY** command on these data. What are the F statistic and its P-value? Give the values of s_p and R^2. Report your overall conclusions about the effect of nematodes on plant growth.

Chapter 11
Inference for Regression

Command to be covered in this chapter:

REGRESS C on K predictors C...C

The REGRESS Command

In Example 11.6 of BPS, the relationship is examined between the number of beers a student drinks and his or her blood alcohol content (BAC). Sixteen student volunteers at Ohio State University drank a randomly assigned number of cans of beer. Thirty minutes later, their BAC was measured. Here and in EG11.06.MTW are the data:

Student	1	2	3	4	5	6	7	8
Beers	5	2	9	8	3	7	3	5
BAC	0.10	0.03	0.19	0.12	0.04	0.095	0.07	0.06

Student	9	10	11	12	13	14	15	16
Beers	3	5	4	6	5	7	1	4
BAC	0.02	0.05	0.07	0.10	0.085	0.09	0.01	0.05

Before attempting inference, examine the data by (1) making a scatterplot, (2) fitting the least squares regression, $\hat{y} = a + bx$, (3) checking for outliers and influential observations, and (4) computing the value of r^2. These can all be done at once by making a fitted line plot. Fitted line plots can be obtained by selecting **Stat ➤ Regression ➤ Fitted line plot** from the menu and entering the appropriate predictor and response variable.

168   Chapter 11

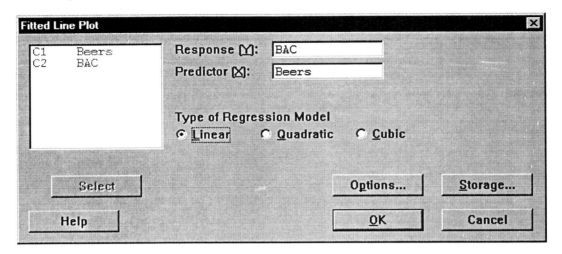

The scatterplot shows a clear linear relationship. The least-squares line is given to be

$$y = -0.013 + 0.018x,$$

and $r^2 = 0.80$. The number of beers you drink predicts blood alcohol level quite well.

Recall that in Chapter 2, the command format for the **REGRESS** command was given as

```
REGRESS C on K predictors C...C
```

We let K = 1, since only one predictor will be used. The **RESIDUALS** subcommand can be used to store the residuals in a separate column using the following format:

```
RESIDUALS put into C
```

The **REGRESS** command and **PREDICT** subcommand can also be used by selecting **Stat ➤ Regression ➤ Regression** from the menu. The Response variable (BAC) and the Predictor variable (Beers) are entered in the dialog box. To store the residuals, clicking on the Storage button and check Residuals in the Regression Storage subdialog box.

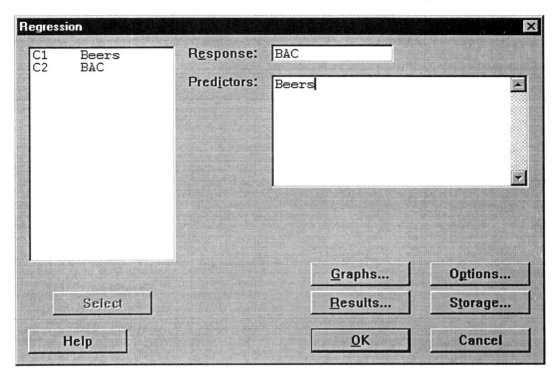

```
MTB > Regress 'BAC' 1 'Beers';
SUBC>    Residuals 'RESI1'.
```

Regression Analysis

The regression equation is
BAC = - 0.0127 + 0.0180 Beers

Predictor	Coef	StDev	T	P
Constant	-0.01270	0.01264	-1.00	0.332
Beers	0.017964	0.002402	7.48	0.000

S = 0.02044   R-Sq = 80.0%   R-Sq(adj) = 78.6%

Analysis of Varianc

Source	DF	SS	MS	F	P
Regression	1	0.023375	0.023375	55.94	0.000
Residual Error	14	0.005850	0.000418		
Total	15	0.029225			

Unusual Observations

Obs	Beers	BAC	Fit	StDev Fit	Residual	St Resid
3	9.00	0.19000	0.14897	0.01128	0.0410	2.41R

R denotes an observation with a large standardized residual

The values of a and b are given in the column labeled Coef. The column labeled Predictor tells us that the first entry, Constant, is a, the intercept, and the second is b, the slope. We see that $a = -0.0127$ and $b = 0.017964$. These are the estimates of α and β. These values are rounded and appear in the regression equation

$$BAC = -0.0127 + 0.0180 \text{ Beers}.$$

The standard error, $s = 0.02044$, is used to estimate σ, the standard deviation of responses about the true regression line.

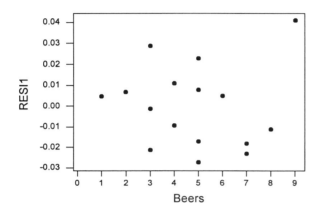

In the example above, we have fitted a line and we should now examine the residuals. Student 3 drank 9 beers and has a residual of .041. This is a mild outlier, but has little effect on r^2 or the fitted line. The assumption of normally distributed residuals appears to be reasonable. This is important for the inference that follows.

Character Stem-and-Leaf Display

```
Stem-and-leaf of RESI1    N  = 16
Leaf Unit = 0.0010

     3    -2  731
     6    -1  871
     8    -0  91
     8     0  4467
     4     1  0
     3     2  28
     1     3
     1     4  1
```

Confidence Intervals and Hypothesis Tests for α and β

Confidence intervals and tests for the slope and intercept are based on the normal sampling distributions of the estimates a and b. Since the standard deviations are not known, a t distribution is used. The value of SE_b appears in the output from the **REGRESS** command to the right of the estimated slope, $b = 0.017964$, it is 0.002402. Similarly, the value of SE_a appears to the right of the estimated con-

Inference for Regression 171

stant, $a = -0.01270$, it is 0.01264. Confidence intervals for α and β have the form

$$\text{estimate} \pm t^* \text{SE}_{\text{estimate}}$$

That is, $\beta = b \pm t^* \text{SE}_b$, where t^* is the upper $(1-C)/2$ critical value for the t distribution with $n-2$ degrees of freedom. The value of t can be calculated using the **INVCDF** command. For a 95% confidence interval, $C = 0.95$.

```
MTB > indf .025;
SUBC> t 14.
      0.0250   -2.1448
```

Therefore, the value of t^* is 2.1448. The upper and lower bounds for the confidence interval can be calculated with Minitab's calculator or using the **LET** command as shown below.

```
MTB > Let k1 = .017964-2.1488*.002402
MTB > Let k2 = .017964+2.1488*.002402
MTB > print k1 k2
```

Data Display

```
K1     0.0128026
K2     0.0231254
```

Therefore, a 95% confidence interval for β is (.0128, .0231). The t statistic and P-value for the test of

$$H_0: \beta = 0$$
$$H_a: \beta \neq 0$$

appear in the columns labeled T and P. The t ratio can also be obtained from the formula

$$t = \frac{b}{\text{SE}_b} = \frac{.017964}{.002402} = 7.48$$

The P-value is listed as 0.000. This means that the P-value is less than 0.001. There is strong evidence against the null hypothesis. Confidence intervals and hypothesis tests for α can be obtained similarly.

The PREDICT Subcommand

We found that the least-squares line for predicting BAC from beer consumption is

$$\hat{y} = -0.0127 + 0.0180x.$$

The **PREDICT** subcommand can be used to predict BAC for students who drink five beers.

```
PREDICT for K
```

When using the menu command, **Stat ➤ Regression ➤ Regression**, click on the Options button. Specify a predicted value in the Regression – Options subdialog box.

```
MTB > regr c2 1 c1;
SUBC> pred 5.
         .
         . (same regression output as above)
         .
   Fit Predicted Values

       Fit   StDev Fit      95.0% CI              95.0% PI
    0.07712    0.00513   ( 0.06612, 0.08812)   ( 0.03192, 0.12232)
```

We may be interested in predicting the *mean* blood alcohol level for *all students* who drink five beers or we mean want to predict the BAC of *one individ-*

ual student who drinks five beers. The prediction is the same for both, $\hat{y} = 0.07712$. However, the margin of error is different for the two kinds of prediction.

Individuals who drink five beers don't all have the same BAC. So, we need a larger margin of error to predict *one individual* BAC than to predict the *mean* BAC for all students who drink five beers. The interval for the *mean* is listed under 95% CI as (0.06612, 0.08812). The interval is $\hat{y} \pm t * SE_{\hat{\mu}}$ and the value labeled Stdev.Fit (0..00513) is $SE_{\hat{\mu}}$. The *individual* prediction interval is listed under 95% PI as (0.03192, 0.12232). This interval is $\hat{y} \pm t * SE_{\hat{y}}$. The value of $SE_{\hat{y}}$ is not given on the Minitab output, but it is easily obtained from the following formula

$$SE_{\hat{y}} = \sqrt{s^2 + (SE_{\hat{\mu}})^2}$$

Remember that before using regression inference, the data must satisfy the regression model assumptions. Use a scatterplot to check that the true relationship is linear. The scatter of the data points about the line should be roughly the same over the entire range of the data. A plot of the residuals against *x* should not show any pattern. A histogram or stemplot of the residuals should not show any major departures from normality.

EXERCISES

11.1 *Archaeopteryx* is an extinct beast having feathers like a bird but teeth and a long bony tail like a reptile. Here are the lengths in centimeters of the femur (a leg bone) and the humerus (a bone in the upper arm) for the five fossil specimens that preserve both bones.

Femur	38	56	59	64	74
Humerus	41	63	70	72	84

The strong linear relationship between the lengths of the two bones helped persuade scientists that all five specimens belong to the same species.

(a) Enter the data into a Minitab worksheet. Make a **PLOT** with femur length as the explanatory variable. Obtain the **CORELATION** and the equation of the least-squares **REGRESSION** line. Do you think that femur length will allow good prediction of humerus length?

(b) Explain in words what the slope β of the true regression line says about *Archaeopteryx*. What is the estimate of β from the data? What is your estimate of the intercept α of the true regression line? What is your estimate of the standard deviation σ in the regression model? You have now estimated all three parameters in the model.

11.2 Below and in TA11-02.MTW are data on the natural gas consumption of the Sanchez household for 16 months. Gas consumption is higher in cold weather. The table gives the average amount of natural gas consumed each day during the month, in hundreds of cubic feet, and the average number of heating degree-days each day during the month. (Degree-days are the number of degrees the average daily temperature falls below 65° F.)

	Degree	Gas		Degree	Gas
Nov.	24	6.3	July	0	1.2
Dec.	51	10.9	Aug.	1	1.2
Jan.	43	8.9	Sept.	6	2.1
Feb.	33	7.5	Oct.	12	3.1
Mar.	26	5.3	Nov.	30	6.4
Apr.	13	4.0	Dec.	32	7.2
May	4	1.7	Jan.	52	11.0
June	0	1.2	Feb.	30	6.9

(a) We want to predict gas used from degree-days. Make a **PLOT** of the data with this goal in mind. Use the **REGRESS** command to find r^2 and the equation of the least-squares regression line. Describe the form and strength of the relationship.

(b) Use the **RESIDUALS** subcommand to find the residuals for all 16 data points. Check that their sum is zero.

(c) The model for regression inference has three parameters, which we call α, β, and σ. Estimate these parameters from the data.

11.4 Use the information in exercise 11.2 to give a 90% confidence interval for the slope β of the true regression line. Explain clearly what your result tells us about how gas usage responds to falling temperatures.

11.5 Here and in EX11-05.MTW are data on the time (in minutes) Professor Moore takes to swim 2000 yards and his pulse rate (beats per minute) after swimming:

Time	34.12	35.72	34.72	34.05	34.13	35.72	36.17	35.57
Pulse	152	124	140	152	146	128	136	144
Time	35.37	35.57	35.43	36.05	34.85	34.70	34.75	33.93
Pulse	148	144	136	124	148	144	140	156
Time	34.60	34.00	34.35	35.62	35.68	35.28	35.97	
Pulse	136	148	148	132	124	132	139	

(a) Make a **PLOT** illustrating the negative linear relationship: A faster time (fewer minutes) is associated with a higher heart rate.

(b) Calculate the **REGRESSION** coefficients and give a 90% confidence interval for the slope of the true regression line. Explain what your result tells us about the relationship between the professor's swimming time and heart rate.

(c) One day the swimmer completes his laps in 34.3 minutes, but forgets to take his pulse. **PREDICT** his heart rate when $x^* = 34.3$. Choose one of the intervals from the output to estimate the swimmer's heart rate that day and explain why you chose this interval.

11.7 There is some evidence that drinking moderate amounts of wine helps prevent heart attacks. Table 2.3 of BPS and EX11-07.MTW give data on yearly wine consumption (liters of alcohol from drinking wine, per person) and yearly deaths from heart disease (deaths per 100,000 people) in 19 developed nations. Use the **REGRESS** command to determine whether there is statistically significant evidence that the correlation between wine consumption and heart disease deaths is negative.

11.8 Exercise 2.6 in BPS and EX11-08.MTW give data on the fuel consumption of a small car at various speeds from 10 to 150 kilometers per hour. Use the **REGRESS** command to determine whether there is evidence of straight-line dependence between speed and fuel use. Make a **PLOT** and use it to explain the result of your test.

11.10 Exercise 11.2 and TA11-02.MTW give the data concerning the natural gas consumed by the Sanchez household on degree-days. In the month of January, after solar panels were installed, there were 40 degree-days per day.

(a) Use the **PREDICT** subcommand to determine how much gas the Sanchez household would have used per day in January without the solar panels. They actually used 75 hundred cubic feet per day. How much gas per day did the solar panels save?

176   Chapter 11

(b) Give a 95% interval for the amount of gas that would have been used this January without the solar panels.

(c) Give a 95% interval for the mean gas consumption per day in months with 40 degree-days per day.

(d) Minitab gives only one of the two standard errors used in prediction. It is $SE_{\hat{\mu}}$. Calculate the other standard error, $SE_{\hat{y}}$.

11.12 Find the **RESIDUALS** for the Sanchez household gas consumption data in TA11-02.MTW.

(a) Display the distribution of the residuals in a **STEM-AND-LEAF** plot. It is hard to assess the shape of a distribution from only 16 observations. Do the residuals appear roughly symmetric? Are there any outliers?

(b) **PLOT** the residuals against the explanatory variable, degree-days. Draw a horizontal line at height zero on the plot. Is there clear evidence of a nonlinear relationship? Does the variation about the line appear roughly the same as the number of degree-days changes?

11.13 Manatees are large, gentle sea creatures that live along the Florida coast. Many manatees are killed or injured by powerboats. Here and in EX11–13.MTW are data on powerboat registrations (in thousands) and the number of manatees killed by boats in Florida in the years 1977 to 1990.

Year	Powerboat registrations	Manatees killed	Year	Powerboat registrations	Manatees killed
1977	447	13	1984	559	35
1978	460	21	1985	585	33
1979	481	24	1986	614	33
1980	498	16	1987	645	39
1981	513	24	1988	675	43
1982	512	20	1989	711	50
1983	526	15	1990	719	47

(a) Make a **PLOT** showing the relationship between powerboats registered and manatees killed. (Which is the explanatory variable?) Is the overall pattern roughly linear? Are there clear outliers or strongly influential data points?

(b) What does R-sq tell you about the relationship between boats and manatees killed?

(c) Explain what the slope β of the true regression line means in this setting. Then give a 90% confidence interval for β.

11.14 The previous exercise and EX11-13.MTW give data on the Florida manatee for the years 1977 to 1990.

(a) Based on these data, you want to **PREDICT** the number of manatees killed in a year when 716,000 powerboats are registered. Use the **REGRESS** command to give a prediction.

(b) Give a 95% interval for the number of manatees that would be killed in a future year if boat registrations remained at 716,000.

(c) Here are four more years of data:

Year	Powerboat registrations	Manatees killed	Year	Powerboat registrations	Manatees killed
1991	716	53	1993	716	35
1992	716	38	1994	735	49

It happens that powerboat registrations remained at 716,000 for the next three years. Did the numbers of manatees killed in those years fall within your prediction interval?

11.16 Ecologists sometimes find rather strange relationships in our environment. One study seems to show that beavers benefit beetles. The researchers laid out 23 circular plots, each four meters in diameter, in an area where beavers were cutting down cottonwood trees. In each plot, they measured the number of stumps from trees cut by beavers and the number of clusters of beetle larvae. Here and in EX11-16.MTW are the data:

Stumps	2	2	1	3	3	4	3	1	2	5	1	3
Beetle larvae	10	30	12	24	36	40	43	11	27	56	18	40
Stumps	2	1	2	2	1	1	4	1	2	1	4	
Beetle larvae	25	8	21	14	16	6	54	9	13	14	50	

(a) Make a **PLOT** that shows how the number of beaver-caused stumps influences the number of beetle larvae clusters. What does your plot show?

(b) Find the least-squares regression line and draw it on your plot. What percent of the observed variation in beetle larvae counts can be explained by straight-line dependence on beaver stump counts?

(c) Is there strong evidence that beaver stumps help explain beetle larvae counts? State hypotheses, give a test statistic and its P-value, and state your conclusion.

178   Chapter 11

11.17 Minitab can calculate *standardized residuals* as well as the actual residuals from regression. Because the standardized residuals have the standard z-score scale, it is easier to judge whether any are extreme. For the data from 11.16, use the **REGRESS** command with the **SRESIDUALS** subcommand or select Standardized Residuals in the Storage subdialog box.

(a) Find the mean and standard deviation of the standardized residuals. Why do you expect values close to those you obtain?

(b) Make a stemplot of the standardized residuals. Are there any striking deviations from normality? The most extreme residual is $z = -1.99$. Would this be surprisingly large if the 23 observations had a normal distribution? Explain your answer.

(c) Plot the standardized residuals against the explanatory variable. Are there any suspicious patterns?

11.18 Investors ask about the relationship between returns on investments in the United States and investments overseas. Here and in EX11-18.MTW are data on the total returns on U.S. and overseas common stocks over a 22-year period. (The total return is change in price plus any dividends paid, converted into U.S. dollars. Both returns are averages over many individual stocks.)

Year	Overseas % return	U.S. % return	Year	Overseas % return	U.S. % return
1971	29.6	14.6	1985	56.2	31.6
1972	36.3	18.9	1986	69.4	18.6
1973	−14.9	−14.8	1987	24.6	5.1
1974	−23.2	−26.4	1988	28.3	16.6
1975	35.4	37.2	1989	10.5	31.5
1976	2.5	23.6	1990	−23.4	−3.1
1977	18.1	−7.4	1991	12.5	30.4
1978	32.6	6.4	1992	−11.8	7.6
1979	4.8	18.2	1993	32.9	10.1
1980	22.6	32.3	1994	6.2	1.3
1981	−2.3	−5.0	1995	11.2	37.6
1982	−1.9	21.5	1996	6.1	23.0
1983	23.7	22.4	1997	2.1	33.4
1984	7.4	6.1			

(a) Make a **PLOT** suitable for predicting overseas returns from U.S. returns.

(b) Find the REGRESSION parameters. What is the value of *t*? What are its degrees of freedom? How strong is the evidence for a linear relationship between U.S. and overseas returns?

(c) PREDICT the overseas returns when U.S. stocks return 15%. You think U.S. stocks will return 15% next year. Give a 90% interval for the return on foreign stocks next year if you are right about U.S. stocks.

(d) Is the regression prediction useful in practice? Use the r^2-value for this regression to help explain your finding.

11.19 Exercise 11.18 presents a REGRESSION of overseas stock returns on U.S. stock returns based on 27 years' data. Find the RESIDUALS for these data.

(a) PLOT the residuals against *x*, the U.S. return. The plot suggests a mild violation of one of the regression assumptions. Which one?

(b) Display the distribution of the residuals in a graph. In what way is the shape somewhat nonnormal? There is one possible outlier. Circle that point on the residual plot in (a). What year is this? This point is not very influential: Redoing the regression without it does not greatly change the results. With 27 observations, we are willing to do regression inference for these data.

11.21 Metabolic rate, the rate at which the body consumes energy, is important in studies of weight gain, dieting, and exercise. Lean body mass is an important influence on metabolic rate. Table 2.2 of BPS and EX11-21.MTW give data for 19 people. Because men and women showed a similar pattern, we will now ignore gender. Here are the data on mass (in kilograms) and metabolic rate (in calories):

Mass	62.0	62.9	36.1	54.6	48.5	42.0	47.4	50.6	42.0	48.7
Rate	1792	1666	995	1425	1396	14.8	1362	1502	1256	1614
Mass	40.3	33.1	51.9	42.4	34.5	51.1	41.2	51.9	46.9	
Rate	1189	913	1460	1124	1052	1347	1204	1867	1439	

Make a PLOT and find the least-squares REGRESSION line. Give a 90% confidence interval for the slope β and explain clearly what your interval says about the relationship between lean body mass and metabolic rate. Find the RESIDUALS and examine them. Are the assumptions for regression inference met?

180   Chapter 11

11.23   How well does a car's city gas mileage (as measured by the Environmental Protection Agency) predict its highway gas mileage? EX11-23.MTW gives the city and highway gas mileages for 29 midsize and 26 small cars from the 1998 model year. The first column specifies the size of the car (1 = midsize, 2 = small), the second column gives the city mileage, and the third gives the highway mileage.

(a)   Make a **PLOT** of the data. Use a different plot symbol for midsize and small cars. Is the association positive or negative? Is there an overall linear pattern? How do the two sizes of cars differ in your plot?

(b)   There is one clear outlier. Circle it in your plot. This is the Volkswagen Passat with diesel engine. Find the correlation between city and highway mileage both with and without this outlier. Why does r decrease when we remove this point? Because this is the only car with a diesel engine, drop it from the data before doing any more work.

(c)   Find the equation of the least-squares **REGRESSION** line for predicting highway mileage from city mileage. Draw the line on your plot from (a).

(d)   Find the **RESIDUALS**. The two most extreme residuals are −6.43 and 4.78. Locate and circle on your plot in (a) the data points that produce these residuals. In what way are these cars unusual? Make a graph of the distribution of the residuals and a plot of the residuals against city mileage. The distribution is reasonably symmetric (with the two possible outliers we have noted) and the residual plot is acceptable.

(e)   Explain the meaning of the slope β of the true regression line in this setting. Give a 95% confidence interval for β.

11.24   Table 11.3 and TA11-02.MTW contain data on the size of perch caught in a lake in Finland.

(a)   We want to know how well we can predict the width of a perch from its length. Make a **PLOT** of width against length. There is a strong linear pattern, as expected. Perch number 143 had six newly eaten fish in its stomach. Find this fish on your scatterplot and circle the point. Is this fish an outlier in your plot of width against length?

(b)   Find the least-squares **REGRESSION** line to predict width from length.

Inference for Regression 181

(c) The length of a typical perch is about $x = 27$ centimeters. **PREDICT** the mean width of such fish and give a 95% confidence interval.

(d) Examine the **RESIDUALS**. Is there any reason to mistrust inference? Does fish number 143 have an unusually large residual?

11.25 We can also use the data in Table 11.3 and TA11-03.MTW to study the prediction of the weight of a perch from its length.

(a) Make a **PLOT** of weight versus length, with length as the explanatory variable. Describe the pattern of the data and any clear outliers.

(b) It is more reasonable to expect the one-third power of the weight to have a straight-line relationship with length than to expect weight itself to have a straight-line relationship with length. Explain why this is true. (Hint: What happens to weight if length, width, and height all double?)

(c) Use Minitab's calculator to create a new variable that is the one-third power of weight. Make a **PLOT** of this new response variable against length. Describe the pattern and any clear outliers.

(d) Is the straight-line pattern in (c) stronger or weaker than that in (a)? Compare the plots and also the values of r^2.

(e) Find the least-squares **REGRESSION** line to predict the new weight variable from length. Predict the mean of the new variable for perch 27 centimeters long, and give a 95% confidence interval.

(f) Examine the **RESIDUALS** from your regressions. Does it appear that any of the regression assumptions are not met?

Chapter 12*
Nonparametric Tests

Commands to be covered in this chapter:

```
MANN-WHITNEY test with [K% confidence] on C C
WTEST [of median = K] on C...C
WINTERVAL [K% confidence] on C...C
KRUSKAL-WALLIS test for data in C, levels in C
```

The MANN-WHITNEY Command

The **MANN-WHITNEY** command does a two-sample rank test (often called the Mann-Whitney test, or the two-sample Wilcoxon rank sum test) for the difference between two population medians, and calculates the corresponding point estimate and confidence interval. The command format is given below.

```
MANN-WHITNEY test with [K% confidence] on C C
```

You can specify a confidence level on the command line. If you do not specify a confidence level, **MANN-WHITNEY** gives a 95% confidence interval. **MANN-WHITNEY** assumes the data are independent random samples from two populations that have the same shape (hence the same variance) and a scale that is at least ordinal. The data need not be from normal populations. The **MANN-WHITNEY** determines the attained significance level of the test using a normal approximation with a continuity correction factor.

The command will be illustrated below on data from Example 12.1 in BPS. A researcher planted corn in 8 plots of ground, then weeded the corn to allow no weeds in 4 plots and exactly 3 weeks per meter in the other 4 plots. Here are the yields of corn (bushels per acre) in each of the plots.

---

* This chapter refers to material available on *The Basic Practice of Statistics* CD-ROM.

Weeds per meter	Yield (bu./acre)			
0	166.7	172.2	165	176.9
3	158.6	176.4	153.1	156.0

First, we will use **STEM-AND-LEAF** plots to examine the shape of the two distributions.

```
MTB > stem c1 c2
```

Character Stem-and-Leaf Display

```
Stem-and-leaf of 0 Weeds    N = 4
Leaf Unit = 1.0

    2     16  56
    2     17  2
    1     17  6

Stem-and-leaf of 3 Weeds    N = 4
Leaf Unit = 1.0

    1     15  3
   (2)    15  68
    1     16
    1     16
    1     17
    1     17  6
```

Stemplots suggest that yields may be lower when weeds are present. Since there is one outlier and small samples, we will use the **MANN-WHITNEY** test that does not require normality. The data for this command needs to be in two columns containing data from two populations. The columns do not need to be the same length. The **MANN-WHITNEY** tests

H_0: median$_1$ = median$_2$

H_a: median$_1$ ≠ median$_2$.

If a one-sided test is required, use the **ALTERNATIVE** subcommand. The two-sample Mann-Whitney test can also be performed by selecting **Stat ➤ Non-parametrics ➤ Mann-Whitney** from the menu. In the First Sample and Second Sample columns, enter the columns with sample data, select the appropriate Alternative and click OK.

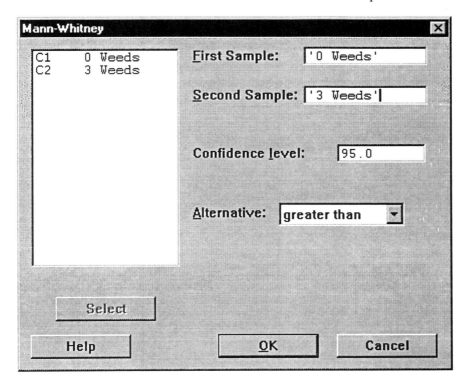

```
MTB > Mann-Whitney 95.0 '0 Weeds' '3 Weeds';
SUBC>    Alternative 1.
```

Mann-Whitney Confidence Interval and Test

```
0 Weeds    N =   4    Median =      169.45
3 Weeds    N =   4    Median =      157.30
Point estimate for ETA1-ETA2 is       11.30
97.0 Percent CI for ETA1-ETA2 is (-11.40,23.80)
W = 23.0
Test of ETA1 = ETA2  vs  ETA1 > ETA2 is significant at 0.0970

Cannot reject at alpha = 0.05
```

The `WTEST` and `WINTERVAL` Command

The `WTEST` command performs a one-sample Wilcoxon signed-rank test of the median. If you do not specify a hypothesized median, `WTEST` compares the sample median to 0. The command format is given below.

```
WILCOXON test [of median = K] on C...C
```

The `WINTERVAL` command calculates a one-sample Wilcoxon confidence interval, separately for the median of each column. The command has the following format.

```
WINTERVAL Wilcoxon CI [K% confidence] on C...C
```

You can specify a confidence level on the command line. If you do not specify a confidence level, `WINTERVAL` gives a 95% confidence interval.

`WINTERVAL` and `WTEST` assume that the data are a random sample from a symmetric population. They do not assume that the population is normal. The Wilcoxon confidence interval and hypothesis test are useful for matched pairs studies such as the Golf scores presented in Example 12.11 of BPS. Here and in EG12-11.MTW are golf scores of 12 members of a college women's golf team in two rounds of tournament play.

Player	1	2	3	4	5	6	7	8	9	10	11	12
Round 2	94	85	89	89	81	76	107	89	87	91	88	80
Round 1	89	90	87	95	86	81	102	105	83	88	91	79
Difference	5	-5	2	-6	-5	-5	5	-16	4	3	-3	1

We will use the `WTEST` command to test the hypothesis that

H_0: scores have the same distribution in rounds 1 and 2

H_a: scores are systematically lower or higher in round 2

The command can either be used in the Session window or by selecting **Stat ➤ Nonparametrics ➤ 1-Sample Wilcoxon** from the menu.

```
MTB > WTest 'Difference'
```

Wilcoxon Signed Rank Test

```
Test of median = 0.000000 versus median not = 0.000000

                 N for  Wilcoxon            Estimated
           N     Test   Statistic    P       Median
Differen   12     12       27.5    0.388     -1.000
```

The output shows that these data give no evidence for a systematic difference in scores between the rounds.

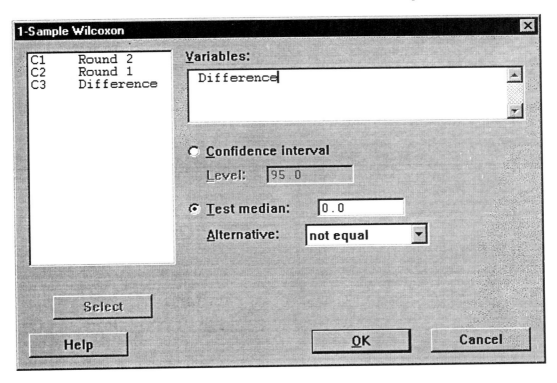

The Wilcoxon confidence interval can also be selected by selecting Confidence interval instead of Test median in the dialog box. Alternatively, the **WINTERVAL** command can be used as shown below. The interval includes zero, consistent with the test above.

```
MTB > WInterval 'Difference'
```

Wilcoxon Signed Rank Confidence Interval

```
              Estimated   Achieved
         N      Median    Confidence  Confidence Interval
Differen 12     -1.00       94.5    (  -5.50,    3.00)
```

The KRUSKAL-WALLIS Command

The **KRUSKAL-WALLIS** command performs a nonparametric alternative to the usual one-way analysis of variance. This test is a generalization of the procedure used by **MANN-WHITNEY**. The test assumes that the data arise as k independent random samples from continuous distributions, all having the same shape (normal or not). The null hypothesis of no differences among the k populations is tested against the alternative of at least one difference. The command format is given below.

KRUSKAL-WALLIS test for data in C, levels in C

The factor column may be numeric or text, and may contain any value. The levels do not need to be in any special order. Example 12.13 provides an example for illustration of the **KRUSKAL-WALLIS** command. In this example, 16 plots were planted with corn. The researched weeded the plots to allow 0, 1, 3, or 9 weeds per meter to grow. Here and in EG12-13.MTW are the yields of corn in each of the plots:

Weeds per meter	Corn yield	Weeds per meter	Corn yield	Weeds per meter	Corn yield	Weeds per meter	Corn yield
0	166.7	1	166.2	3	158.6	9	162.8
0	172.2	1	157.3	3	176.4	9	142.4
0	165.0	1	166.7	3	153.1	9	162.7
0	176.9	1	161.1	3	156.0	9	162.4

The data is entered in a Minitab worksheet with the number of weeds per meter in C1 and the Corn yield in C2.

```
MTB > info
```

Information on the Worksheet

```
Column   Count   Name
C1        16     Weeds
C2        16     Yield

MTB > desc c2;
SUBC> by c1.
```

Descriptive Statistics

```
Variable   Weeds         N      Mean    Median    TrMean     StDev
Yield        0           4    170.20    169.45    170.20      5.42
             1           4    162.82    163.65    162.82      4.47
             3           4    161.03    157.30    161.03     10.49
             9           4    157.58    162.55    157.58     10.12

Variable   Weeds    SE Mean   Minimum   Maximum        Q1        Q3
Yield        0         2.71    165.00    176.90    165.43    175.73
             1         2.23    157.30    166.70    158.25    166.57
             3         5.25    153.10    176.40    153.82    171.95
             9         5.06    142.40    162.80    147.40    162.78
```

The sample standard deviations do not satisfy our rule of thumb for the safe use of ANOVA and the yields for 3 and 9 weeds per meter have outliers. Therefore it is reasonable to use a nonparametric test.

The `KRUSKAL-WALLIS` command can be used in the session window or by selecting **Stat ➤ Nonparametrics ➤ Kruskal-Wallis** from the menu and filling in the columns for Response and Factor.

```
MTB > Kruskal-Wallis 'Yield' 'Weeds'.
```

Kruskal-Wallis Test

```
Kruskal-Wallis Test on Yield

Weeds        N    Median    Ave Rank         Z
0            4     169.4        13.1      2.24
1            4     163.6         8.4     -0.06
3            4     157.3         6.2     -1.09
9            4     162.6         6.2     -1.09
Overall     16                   8.5

H = 5.56   DF = 3   P = 0.135
H = 5.57   DF = 3   P = 0.134  (adjusted for ties)

* NOTE * One or more small samples
```

The *P*-value is equal to .135 (.134 adjusted for ties) which does not provide convincing evidence that weeks have an effect.

EXERCISES

12.1 A study of early childhood education asked kindergarten students to tell a fairy tale that had been read to them earlier in the week. The 10 children in the study included 5 high-progress readers and 5 low-progress readers. Each child told two stories. Story 1 had been read to them; Story 2 had been read and also illustrated with pictures. An expert listened to a recording of the children and assigned a score for certain uses of language. Is there evidence that the scores of high-progress readers are higher than those of low-progress readers when they retell a story they have heard without pictures (Story 1)? The data are in EX12-01.MTW.

    (a) Make `STEM-AND-LEAF` plots for the 5 Story 1 scores in each group. Are any major deviations from normality apparent?

    (b) Use the `TWOT` command to carry out a two-sample *t* test. State hypotheses and give the *P*-value and your conclusion.

    (c) `UNSTACK` the Story 1 scores into 2 columns (high and low). Use the `MANN-WHITNEY` command to carry out the Wilcoxon rank sum test. State hypotheses and give the *P*-value and your conclu-

190   Chapter 12

sion. Do the *t* and Wilcoxon tests lead you to different conclusions?

(d) Repeat (a), (b), and (c) for the scores when children retell a story they have heard and seen illustrated with pictures (Story 2).

12.6 In Example 7.8, we compared the breaking strength of polyester strips buried for 16 weeks with that of strips buried for 2 weeks. The breaking strength in pounds were:

2 weeks	118	126	126	120	129
16 weeks	124	98	110	140	110

(a) Use the **MANN-WHITNEY** command to test to these data and compare your result with the $P = 0.19$ obtained from the two-sample *t* test in Example 7.7.

(b) What are the null and alternative hypotheses for the *t* test? For the Mann-Whitney test?

12.7 Exercise 7.32 of BPS compared the number of tree species in unlogged plots in the rain forest of Borneo with the number of species in plots logged 8 years earlier. Here are the data:

Unlogged	22	18	22	20	15	21	13	13	19	13	19	15
Logged	17	4	18	14	18	14	14	10	12			

(a) Make **STEM-AND-LEAF** plots of the data using the **BY** subcommand. Does there appear to be a difference in species counts for logged and unlogged plots?

(b) Does logging significantly reduce the mean number of species in a plot after 8 years? State the hypotheses, **UNSTACK** the data into 2 columns and do a **MANN-WHITNEY** test. State your conclusion.

12.9 Data in EG12-06.MTW concern a study of the attitudes of people attending outdoor fairs about the safety of the food served at such locations. It contains the responses of 303 people to several questions. The variable "srest" contains responses to the question "How often do you think people become sick because of food served at restaurants?" The responses range from 1 = very rarely to 5 = always. Are women more concerned than men about the safety of food served in restaurants? **UNSTACK** the data from the column names "srest" into separate columns for women and men. Use the **MANN-WHITNEY** command and state your conclusion.

12.15 Example 12.6 of BPS describes a study of the attitudes of people attending outdoor fairs about the safety of the food served at such locations. The data are given in EG12-06.MTW. The variable "srest" contains responses to the safety question (How often do you think people become sick from ...?) asked about food served in restaurants. The responses range from 1 = very rarely to 5 = always. The variable "sfair" contains responses from the same people to the same question asked about food served outdoors at a fair. We suspect that restaurant food will appear safer than food served outdoors at a fair. Do the data give good evidence for this suspicion? Use the **WILCOXON** command. Give the test statistic and its P-value. State your conclusion.

12.17 Table 7.4 in BPS and TA07-04.MTW give data on the healing rate (micrometers per hour) of the skin of newts under two conditions. This is a matched pairs design, with the body's natural electric field for one limb (control) and half the natural value for another limb of the same newt (experimental). We want to know whether the healing rates are systematically different under the two conditions. You decide to use a **WILCOXON** test. State hypotheses, carry out a test, and give your conclusion. Be sure to include a description of what the data show in addition to the test results.

12.18 Cola makers test new recipes for loss of sweetness during storage. Trained tasters rate the sweetness before and after storage. Here are the sweetness losses found by 10 tasters for one new cola recipe. Are these data good evidence that the cola lost sweetness?

   2.0   0.4   0.7   2.0   −0.4   2.2   −1.3   1.2   1.1   2.3

   (a) These data are the differences from a matched pairs design. State hypotheses in terms of the median difference in the population of all tasters, carry out a **WILCOXON** test, and give your conclusion.

   (b) In Example 7.2 we found that the one-sample t test had P-value = 0.012 for these data. How does this compare with your result from (a)? What are the hypotheses for the t test? What assumptions must we make for each of the t and Wilcoxon tests?

12.20 Table 7.2 in BPS and EX12-20.MTW contain data from a student project that investigated whether right-handed people can turn a knob faster clockwise than they can counterclockwise. Describe what the data show, then state hypotheses and do a **WILCOXON** test that does not require normality. Report your conclusions carefully.

12.21 How do nematodes (microscopic worms) affect plant growth? A botanist prepares 16 identical planting pots and then introduces different numbers of nematodes into the pots. A tomato seedling is transplanted into each plot. Here are data on the increase in height of the seedlings (in centimeters) 16 days after planting.

Nematodes	Seedling growth			
0	10.8	9.1	13.5	9.2
1000	11.1	11.1	8.2	11.3
5000	5.4	4.6	7.4	5.0
10,000	5.8	5.3	3.2	7.5

We applied ANOVA to these data in Exercise 10.17. Because the samples are very small, it is difficult to assess normality.

(a) What hypotheses does ANOVA test? What hypotheses does Kruskal-Wallis test?

(b) Find the median growth in each group. Do nematodes appear to retard growth? Apply the **KRUSKAL-WALLIS** test. What do you conclude?

12.22 Example 10.6 in BPS used ANOVA to analyze the results of a study to see which of four colors best attracts cereal leaf beetles. Here and in EX12-22.MTW are the data:

Color	Insects trapped					
Lemon yellow	45	59	48	46	38	47
White	21	12	14	17	13	17
Green	37	32	15	25	39	41
Blue	16	11	20	21	14	7

Because the samples are small, we will apply a nonparametric test.

(a) What hypotheses does ANOVA test? What hypotheses does Kruskal-Wallis test?

(b) Use the **KRUSKAL-WALLIS** test to see whether there are significant differences among the colors. Which colors appear more effective? What do you conclude?

Nonparametric Tests 193

12.23 Table 12.1 and TA12-01.MTW present data on the calorie and sodium content of selected brands of beef, meat, and poultry hot dogs. We will regard these brands as random samples from all brands available in food stores.

(a) Make **STEM-AND-LEAF** plots of the calorie contents. Use the **INCREMENT** subcommand to keep the same scale for easy comparison. Use the **DESCRIBE** command to find the five-number summaries for the three types of hot dog. What do the data suggest about the calorie content of different types of hot dog?

(b) Are any of the three distributions clearly not normal? Which ones, and why?

(c) Apply the **KRUSKAL-WALLIS** test. Report your conclusions carefully.

12.25 Repeat the analysis of Exercise 12.23 for the sodium content of hotdogs.

12.26 Here and in EX12-26.MTW are the breaking strengths (in pounds) of strips of polyester fabric buried in the ground for several lengths of time:

2 weeks	118	126	126	120	129
4 weeks	130	120	114	126	128
8 weeks	122	136	128	146	140
16 weeks	124	98	110	140	110

Breaking strength is a good measure of the extent to which the fabric has decayed.

(a) Use the **DESCRIBE** command to find the standard deviations of the 4 samples. They do not meet our rule of thumb for applying ANOVA. In addition, the sample buried for 16 weeks contains an outlier. We will use the Kruskal-Wallis test.

(b) What are the hypotheses for the Kruskal-Wallis test, expressed in terms of medians?

(c) Carry out the **KRUSKAL-WALLIS** test and report your conclusion.

12.28 Table 10.3 in BPS and TA10-03.MTW contains data comparing the number of trees and number of tree species in plots of land in a tropical rainforest that had never been logged with similar plots nearby that had been logged 1 year earlier and 8 years earlier.

(a) Use side by side **BOXPLOTS** to compare the distributions of number of trees per plot for the three groups of plots. Are there features that might prevent use of ANOVA?

(b) Use the **KRUSKAL-WALLIS** test to compare the distributions of tree counts. State hypotheses, the test statistic and its P-value, and your conclusions.

Appendix

Minitab Session Commands and Menu Equivalents

Notation:

K   denotes a constant such as 8.3 or k14
C   denotes a column, such as C12 or 'Height'
E   denotes either a constant or column
[ ]   encloses an optional argument

1. General Information

HELP `command`
    Help ➤ Search for Help on

INFO `[C...C]`
    Window ➤ Info

STOP
    File ➤ Exit

2. Input and Output of Data

SET `data into C`
    Editor ➤ Insert Columns

INSERT `data [between rows K and K] of C...C`
    Editor ➤ Insert Cells

END `of data`
    not available

NAME `E = 'name' ... E = 'name'`
    In menu versions, use the data editor to name columns

PRINT `the data in E...E`
    not available

SAVE `[in file in "filename" or K]`
    File ➤ Save Worksheet (As)

RETRIEVE `[file in "filename" or K]`
    File ➤ Open Worksheet

3. Editing and Manipulating Data

LET C(K) = K
    not available

DELETE rows K...K of C...C
    Manip ➤ Delete Rows

ERASE E...E
    Manip ➤ Erase Variables

INSERT data [between rows K and K] of C...C
    Editor➤ Insert Cells

STACK (E...E) on ... on (E...E), put in (C...C)
    Manip ➤ Stack/Unstack ➤ Stack Columns

UNSTACK (C...C) into (E...E) ... (E...E)
    Manip ➤ Stack/Unstack ➤ Unstack One Column

SORT C [carry along C...C] put into C [and C...C]
    Manip ➤ Sort

4. Arithmetic

Arithmetic Operations:
    Calc ➤ Calculator

LET = expression

ADD E to E...E, put into E

SUBTRACT E from E, put into E

MULTIPLY E by E...E, put into E

DIVIDE E by E, put into E

RAISE E to the power E put into E

ABSOLUTE value of E put into E

SQRT of E put into E

LOGE of E put into E

LOGTEN of E put into E

EXPONENTIATE E put into E

ANTILOG of E put into E

ROUND E put into E

Columnwise Statistics:
    Calc ➤ Column Statistics

```
COUNT the number of values in C [put into K]
N count the nonmissing values in C [put into K]
NMISS (number of missing values in) C [put into K]
SUM of the values in C [put into K]
MEAN of the values in C [put into K]
MEDIAN of the values in C [put into K]
MINIMUM of the values in C [put into K]
MAXIMUM of the values in C [put into K]
```

Rowwise Statistics:
    Calc ➤ Row Statistics

```
RCOUNT of E...E put into C
RN of E...E put into C
RNMISS of E...E put into C
RSUM of E...E put into C
RMEAN of E...E put into C
RSTDEV of E...E put into C
RMEDIAN of E...E put into C
RMINIMUM of E...E put into C
RMAXMUM of E...E put into C
```

5. Plotting Data

```
HISTOGRAM of C...C
```
    Graph ➤ Histogram

```
STEM-AND-LEAF of C...C
```
    Graph ➤ Stem-and-Leaf

```
BOXPLOT of C...C
```
    Graph ➤ Boxplot

```
PLOT C * C
```
    Graph ➤ Scatter Plot

```
TSPLOT [period = K] of C
```
    Graph ➤ Time Series Plot

6. Basic Statistics

DESCRIBE `variables in C...C`
    Stat ➤ Basic Statistics ➤ Descriptive Statistics

ZINTERVAL `[K% confidence], sigma = K for C...C`
    Stat ➤ Basic Statistics ➤ 1-Sample Z

ZTEST `[of µ = K] assumed sigma = K on C...C`
    Stat ➤ Basic Statistics ➤ 1-Sample Z

TINTERVAL `[K% confidence] for data in C...C`
    Stat ➤ Basic Statistics ➤ 1-Sample t

TTEST `[of µ = K] on data in C...C`
    Stat ➤ Basic Statistics ➤ 1-Sample t

TWOSAMPLE `[K% confidence] for data in C and C`
    Stat ➤ Basic Statistics ➤ 2-Sample t

TWOT `[K% confidence] for data in C, subscripts in C`
    Stat ➤ Basic Statistics ➤ 2-Sample t

PAIR `C C`
    Stat ➤ Basic Statistics ➤ Paired t

CORRELATION `between C...C`
    Stat ➤ Basic Statistics ➤ Correlation

CENTER `the data in C...C put into C...C`
    Calc ➤ Standardize

7. Regression

REGRESS `C on K predictors C...C`
    Stat ➤ Regression ➤ Regression

8. Analysis of Variance

AOVONEWAY `for samples in C...C`
    Stat ➤ ANOVA ➤ Oneway (unstacked)

ONEWAY `data in C, levels in C`
    Stat ➤ ANOVA ➤ Oneway

9. Tables

TALLY `the data in C...C`

Stat ➤ Tables ➤ Tally

`TABLE the data classified C...C`
Stat ➤ Tables ➤ Cross Tabulation

`CHISQUARE test on table stored in C...C`
Stat ➤ Tables ➤ Chisquare Test

10. Statistical Process Control

`ICHART for C`
Stat ➤ Control Charts ➤ Individuals

`XBARCHART [C E]`
Stat ➤ Control Charts ➤ Xbar

`SCHART [C E]`
Stat ➤ Control Charts ➤ S

11. Distributions and Random Data

`RANDOM K observations into C...C`
Calc ➤ Random Data

`PDF for values in E...E [put results in E...E]`
Calc ➤ Probability Distributions

`CDF for values in E...E [put results in E...E]`
Calc ➤ Probability Distributions

`INVCDF for values in E [put into E]`
Calc ➤ Probability Distributions

`SAMPLE K rows from C...C put into C...C`
Calc ➤ Random Data ➤ Sample From Columns

12. Symbols

*   Missing Value Symbol. An asterisk (*) can be used as data in `SET` and `INSERT` and in datafiles. Enclose the asterisk in single quotes in commands and subcommands.

\#   Comment Symbol. The pound sign (#) anywhere on a line tells Minitab to ignore the rest of the line.

&   Continuation Symbol. To continue a command onto another line, end the first line with the ampersand (&).

Index

A
abbreviate, 69, 74
ABSOLUTE, 196
ADD, 196
ALL, 86
analysis of variance (ANOVA), 159–163, 187, 189
ANTILOG, 196
AOVONEWAY, 159, 162–163, 198
arithmetic, 9, 31–32, 90, 196
asterisk (*), 26, 199

B
BERNOULLI, 85, 101
BINOMIAL, 102, 103, 105
BOXPLOT, 13, 29–30, 129, 160, 197
BY, 21, 23, 26

C
CDF, 13, 33, 101, 103, 106, 199
CENTER, 13, 30, 198
center line, 93, 94, 100
central limit theorem, 91
changing the data, 8
CHISQUARE, 151–153, 199
columns, 2, 3, 5, 6, 7, 8, 9, 16, 20, 21, 24, 27, 48, 56, 76, 86, 90, 128, 151, 152, 162, 184, 195
Comment Symbol (#), 199
confidence interval, 111–112, 121–122, 126–128, 139–142, 162, 171, 183, 186–187
constants, 2, 3, 6, 9
Continuation Symbol, 199
control charts, 92–96
control limits, 93
CORRELATION, 48, 49, 198

COUNT, 27, 197
critical value, 114, 124, 130, 171
CUMCOUNTS, 86
CUMPERCENTS, 86

D
data prompt (DATA>), 5
Data window, 2, 4, 8, 9, 16, 88, 153
degrees of freedom, 124, 128, 152, 161–162, 171
DELETE, 8, 9, 196
DESCRIBE, 13, 26, 105, 112, 122, 160, 198
DISCRETE, 88
DIVIDE, 196

E
Editor menu, 4, 9
END, 5, 195
Entering Data, 1, 4–5
equal variances, 128, 129
ERASE, 8, 9, 196
EXPONENTIATE, 32, 196

F
F statistic, 129–130, 161
F Test for Equality of Variance, 129–130

H
Help, 1, 10, 195
HELP, 10, 19, 55, 195
HISTOGRAM, 4, 13, 15–17, 20, 105, 129, 197
histograms, 15–17, 91
History window, 2

hypothesis test, 113–114, 125, 126–128, 139–142, 186–189

I
`ICHART`, 98, 199
`INCREMENT`, 20, 193
`INFO`, 6–7, 10, 195
Info window, 2, 6
`INSERT`, 8, 9, 195, 199
`INVCDF`, 13, 33–34, 111, 114, 124, 130, 171, 199

L
least squares, 167
`LET`, 8, 9, 13, 31–32, 34, 51, 106, 129, 171, 195, 196
`LOGE`, 31, 196
`LOGTEN`, 31, 196
lower control limit, 93, 95

M
Macintosh computers, 2
matched pairs, 124–125, 186
`MAXIMUM`, 27, 28, 31 197
`MEAN`, 13, 27, 28, 31, 90, 197
`MEDIAN`, 27, 28, 197
menu command, 10, 16, 25, 46
`MINIMUM`, 27, 28, 31, 197
Minitab prompt (MTB >), 2, 4
Minitab session, 2
missing values, 26, 27, 86, 197
mu, 33, 89, 122
`MULTIPLY`, 196

N
`N`, 27, 197
`NAME`, 6, 195
`NMISS`, 27, 197
normal, 29–34, 85, 90, 91, 106, 111, 113, 114, 129, 140, 141, 160, 170, 186, 187

`NORMAL`, 33, 34, 89

O
`ONEWAY`, 159–162, 198
out of control signals, 96

P
`PAIR`, 121, 125, 198
`PERCENTS`, 86–87
`PLOT`, 4, 45–47, 53, 197
`POOLED`, 128, 142–143, 162
`PREDICT`, 53, 171–172
`PRINT`, 8, 195
P-value, 113, 123, 125, 127, 130, 152, 162, 171, 189

Q
quartile, 26, 28

R
`RAISE`, 196
`RANDOM`, 85–91, 101–102, 105, 199
random numbers, 85, 89, 101
`RCOUNT`, 28, 197
`REGRESS`, 45, 49–53, 167–170, 198
`RETRIEVE`, 195
`RMAXIMUM`, 28, 197
`RMEAN`, 13, 28, 85, 90, 197
`RMEDIAN`, 28, 197
`RMINIMUM`, 28, 197
`RN`, 28, 197
`RNMISS`, 28, 197
`ROUND`, 31, 196
rows, 2, 5, 8, 9, 21, 27–28, 69, 76, 90, 92, 197
`RSTDEV`, 28, 197
`RSUM`, 28, 197

S
`SAMPLE`, 69–79, 199
`SAVE`, 8, 195

scatterplot, 46–47, 49, 168,
SCHART, 85, 95, 96, 199
SE Mean, 26, 27, 122
Session commands, 4
Session window, 2, 3, 4, 6, 16, 186
SET, 5, 69, 195, 199
side-by-side boxplots, 29, 160
sigma, 33, 89, 111, 113, 198
SORT, 69, 71, 72, 195
SQRT, 31, 196
STACK, 13, 21, 195
standard deviation, 26, 30, 31, 34, 85, 89, 91, 93, 94, 111, 160, 162, 170
standard error, 26, 170
starting a Minitab session, 2
STEM-AND-LEAF, 13, 14, 16, 20, 23, 197
stemplots, 13, 20,
STOP, 2, 3, 195
Stored constants, 2
subcommand prompt (SUBC>), 19
subcommands, 19, 53, 55, 85, 86, 93, 94, 140, 143, 199
SUBSCRIPTS, 21, 24, 76
SUBTRACT, 196
SUM, 28, 197

T
t statistic, 171
t test, 122–128
TABLE, 45, 54–56, 199
TALLY, 85, 86–87, 198
time series, 24–25
TINTERVAL, 121-122, 123, 197
trimmed mean, 26
TSPLOT, 13, 24–25, 197
TTEST, 121, 122–123, 198
TWOSAMPLE, 121, 128, 197
Two-Sample t, 126–128
TWOT, 121, 128, 198

U
UNIFORM, 89
UNSTACK, 13, 23–24, 69, 77, 196
upper control limit, 93

V
versions of Minitab, 2, 4

W
worksheet, 2, 4–6, 7, 8, 16, 73, 74, 76, 79, 104, 112, 126, 151, 188

X
XBARCHART, 85, 92–96, 199

Z
ZINTERVAL, 111–112, 198
ZTEST, 111, 113, 198

Maame 401 221 5636